Electrical and Electronic Engineering

Electrical and Electronic Engineering

Zain Jackson

Willford Press
www.willfordpress.com

Published by Willford Press,
118-35 Queens Blvd., Suite 400,
Forest Hills, NY 11375, USA

ISBN: 978-1-64728-502-9

Cataloging-in-Publication Data

Electrical and electronic engineering / Zain Jackson.
p. cm.
Includes bibliographical references and index.
ISBN 978-1-64728-502-9
1. Electronics. 2. Industrial electronics. 3. Electronic apparatus and appliances.
4. Engineering. I. Jackson, Zain.

TK7835 .P73 2023
621.381--dc23

For information on all Willford Press publications visit our website at www.willfordpress.com

Willford Press

Contents

Preface

Electrical and electronics engineering (EEE) is an integrated branch of engineering that studies the practical applications of electricity in all its forms, including those related to the field of electronics. This branch deals with the technological aspects of electricity, especially the design and application of circuitry and electronic equipment. The concept of power generation and distribution, communication, and machine control are also included in the domain of EEE. It incorporates fundamental knowledge in principal disciplines such as control systems, communications, signal processing, micro-processors, radio frequency design, electric machines, and power generation. It involves the study of various concepts associated with semiconductors, analog electronics, power systems, big transmission lines, digital electronics, electrical machines (AC and DC), and control systems. This book provides a comprehensive coverage of key concepts and models essential to the understanding of electrical and electronic engineering. It will serve as valuable source of reference for scholars and engineers actively engaged in this field.

This book unites the global concepts and researches in an organized manner for a comprehensive understanding of the subject. It is a ripe text for all researchers, students, scientists or anyone else who is interested in acquiring a better knowledge of this dynamic field.

I extend my sincere thanks to the contributors for such eloquent research chapters. Finally, I thank my family for being a source of support and help.

Zain Jackson

1

DC Network Theorem

1.1 Definition of Electric Circuit and Network

An electric circuit or electric network is the interconnection of electrical elements linked together in the closed path so that an electric current may continuously flow. Also, an electric circuit is essentially a pipeline which facilitates the transfer of charge from one point to another.

The interconnection of various electric elements in a prescribed manner comprises an electric circuit in order to perform the desired function. The electric elements include controlled and uncontrolled sources of energy such as resistors, capacitors, inductors, etc. Analysis of electric circuits denote the computations required to determine the unknown quantities such as current, voltage and power associated with one or more elements in the circuit.

Electrical Network

The combination of various electric elements connected in any manner is called an electrical network. We may classify the circuit element into two categories known as active and passive elements.

1.1.1 Linear Circuit and Non-Linear Circuit

Linear Circuit

A linear circuit is one in which the values of the circuit components, (i.e.) resistance, capacitance, inductance, gain, etc., do not change with the level of voltage or current in the circuit. A linear circuit obeys the principle of superposition.

The principle of superposition states that the output of the circuit F(x) when linear combination of signals $ax_1(t) + bx_2(t)$ is applied to it, must be equal to the linear combination of the outputs due to the signals $x_1(t)$ and $x_2(t)$ applied separately.

Since the linear circuits obey the principle of superposition, they can be analyzed with frequency domain mathematical techniques such as, Fourier analysis and Laplace transform. Examples of linear circuits are differentiators, ideal amplifiers, integrators or any circuit made of ideal resistors, capacitors, inductors and op-amps in the unsaturated region.

Non-Linear Circuit

A circuit whose parameters change with the application of voltage or current is called a non-linear circuit. Examples of non-linear behaviour of circuit elements are transistors in saturated region, diodes, saturated iron core inductors and transformers, modulators and digital logic circuits. Non-linear circuits are analysed using approximate numerical methods by electronic circuit simulation computer programs such as spice, for accurate results.

1.1.2 Bilateral Circuit and Unilateral Circuit

Unilateral Circuits

In unilateral circuits, the property of circuit changes with the change of direction of supply voltage or current. In other words, unilateral circuit allows the current to flow in one direction only. Example of unilateral circuit is a diode rectifier, because it does not perform the rectification in both the directions of the supply.

Example: Vacuum tubes, silicon diodes, metal rectifiers, etc.

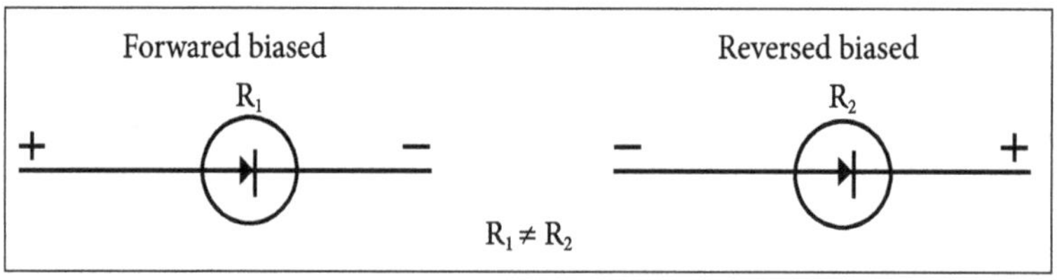

Forward and reversed biased unilateral circuits.

Bilateral Circuits

In bilateral circuits, the property of circuit does not change with the change of direction of supply voltage or current. In other words, bilateral circuit allows the current to flow in both the directions. A best example of bilateral circuit is the transmission line, because if we give supply from any direction, the circuit properties remain constant.

Example: Resistors, inductors and capacitors.

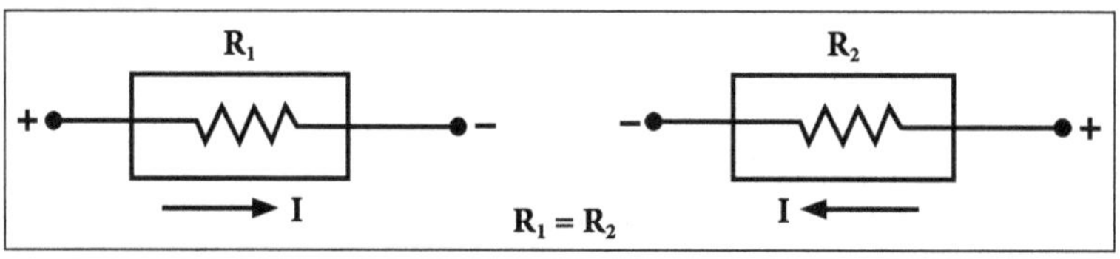

Bilateral circuits.

1.1.3 Energy Sources

Ideal Independent and Dependent Sources

An active two-terminal element that supplies energy to the circuit is a source of energy.

An ideal voltage source is a circuit element which maintains the prescribed voltage across the terminals regardless of the current flowing in those terminals. Similarly, an ideal current source is the circuit element that maintains the prescribed current through its terminals regardless of the voltage across those terminals. These circuit elements do not exist as practical devices, they are only idealized models of actual current and voltage sources.

The ideal voltage and current sources can be further classified as either independent sources or dependent sources. The independent source establishes the current or voltage in a circuit without relying on the currents or voltages elsewhere in the circuit. The value of the current or voltage supplied is specified by the value of the independent source alone.

The dependent source establishes the current or voltage whose value depends on the value of the voltage or current elsewhere in the circuit. We cannot specify the value of the dependent source, unless we know the value of current or voltage on which it depends.

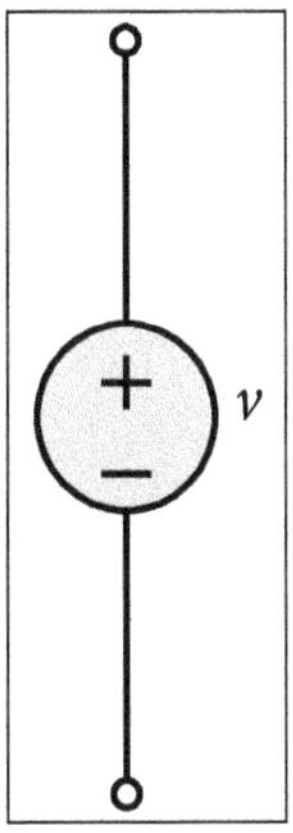

(a) An ideal independent voltage source.

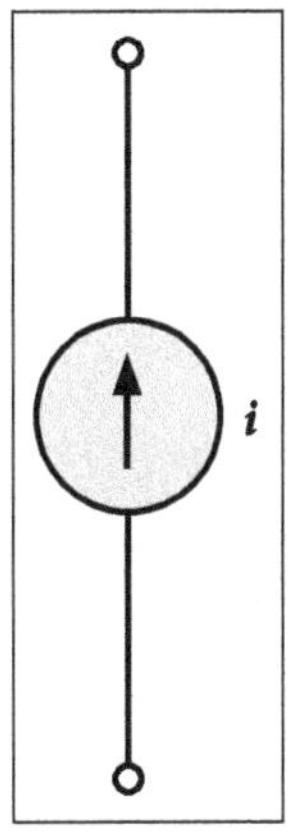

(b) An ideal independent current source.

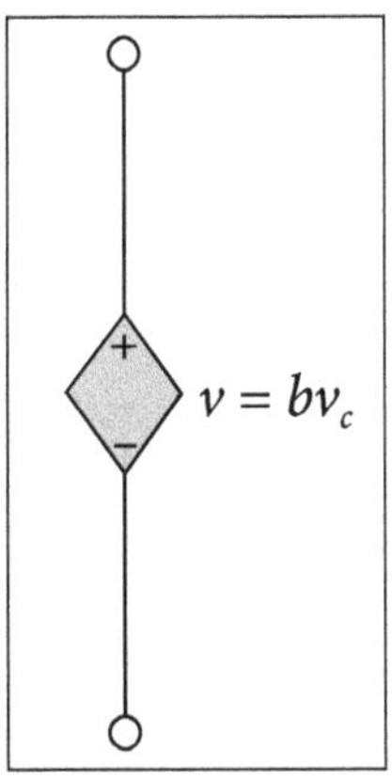

(c) Voltage controlled voltage source.

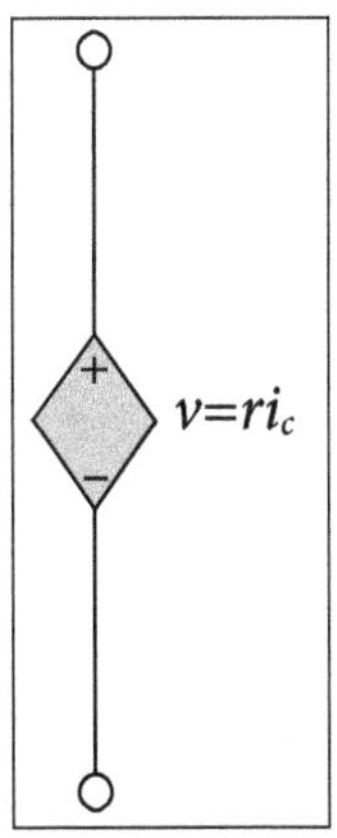

(d) Current controlled voltage source.

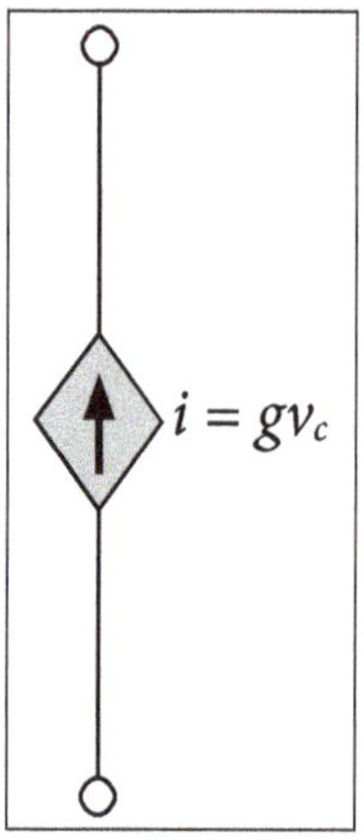

(e) Voltage controlled current source.

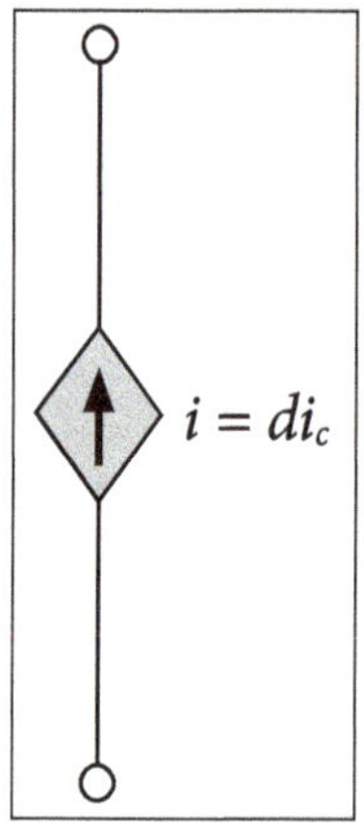

(f) Current controlled current source.

The circuit symbols for ideal independent sources are shown in figure (a) and (b). Circle is used to represent the independent source. The circuit symbols for dependent sources are shown in figure (c), (d), (e) and (f). Diamond symbol is used to represent a dependent source.

Non-Ideal Voltage Sources

The non-ideal voltage source delivers a voltage V and a current i, where:

$$V = V_s - i.R_s \qquad ...(1)$$

Equation (1) indicates that the voltage delivered by the non-ideal voltage source model decreases as the current out of the voltage source increases, which agrees with the expectations.

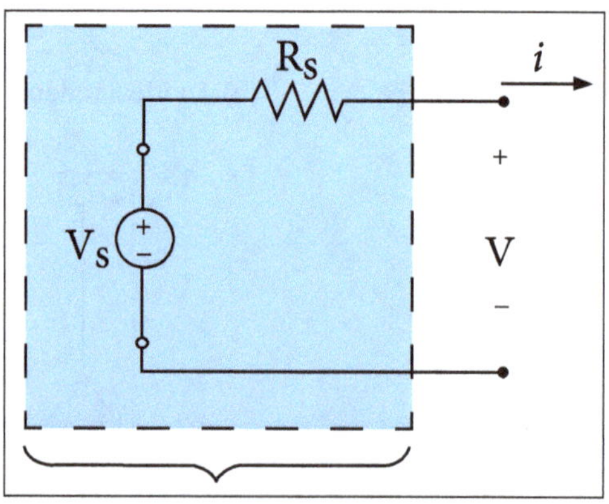

Non-ideal voltage source model.

Non-Ideal Current Sources

An ideal current source was providing a specified current, regardless of the voltage difference across the device. This model suffers from the same basic drawback as our ideal voltage source model. The model can deliver infinite power, which is inconsistent with the capabilities of a real current source. The source delivers a voltage V and current i. The output current is given by:

$$i = i_s - \frac{V}{R_S} \quad ...(2)$$

Equation (2) shows that the current delivered by the source decreases as the delivered voltage increases.

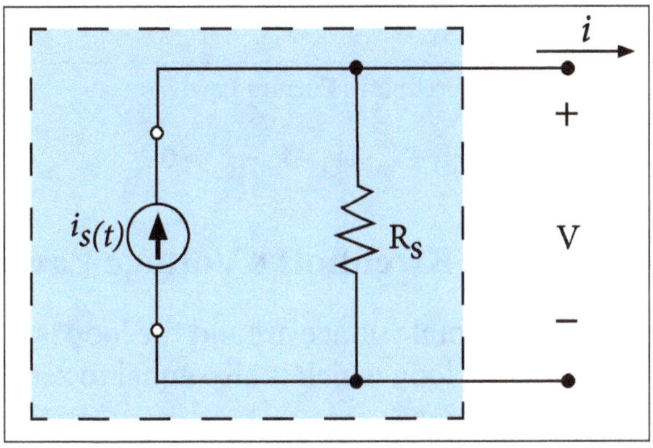

Non-ideal current source model.

1.2 Kirchhoff's Law

Kirchhoff's laws are the basic analytical tools in order to obtain the solutions of voltages and currents for any electric circuit, whether it is supplied from a direct current system or an alternating current system.

Kirchhoff's First Law or Kirchhoff's Current Law (KCL)

"The total charge or current entering a junction or node is exactly equal to the charge leaving the node as it has no other place to go except to leave, as no charge is lost within the node."

In other words, KCL states that at any node (junction) in a circuit, the algebraic sum of currents entering and leaving a node at any instant of time must be equal to zero.

$$I(\text{exiting}) + I(\text{entering}) = 0.$$

This idea by Kirchhoff is known as the Conservation of Charge.

At any instant of time, the algebraic sum of current at a junction or node is zero. Otherwise, the algebraic sum of current entering into a junction is equal to the current leaving the junction.

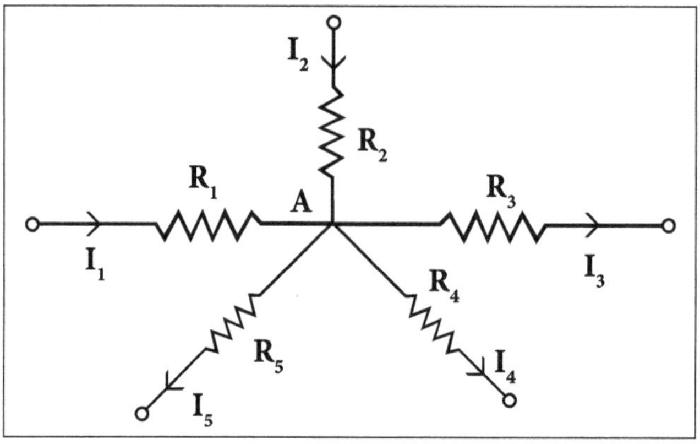

Kirchhoff's Current Law.

$$\Rightarrow I_1 + I_2 = I_3 + I_4 + I_5 \text{ (or) } I_1 + I_2 - I_3 - I_4 - I_5 = 0$$

Kirchhoff's Second Law or Kirchhoff's Voltage Law (KVL)

"In any closed loop network, the total voltage around the loop is equal to the sum of all the voltage drops within the same loop which is also equal to zero."

In other words, KVL states that in a closed circuit, the algebraic sum of all the source voltages must be equal to the algebraic sum of all the voltage drops. This idea by Kirchhoff is known as the Conservation of Energy.

The algebraic sum of potential around a closed path is equal to zero. In other words, the sum of potential rise is equal to the sum of potential drop in a closed circuit.

$$V = V_1 + V_2 + V_3 \text{ (or) } V - V_1 - V_2 - V_3 = 0$$

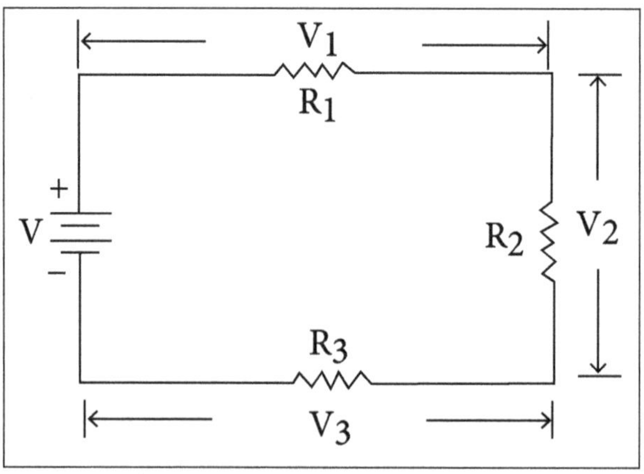

Kirchhoff's Voltage Law.

Steps to solve a problem using KCL/KVL method:

- Step 1: Find the branch currents of each branch using KCL.

- Step 2: Using KVL, get the equations for each loop in terms of unknown currents.
- Step 3: Solve the simultaneous equations obtained in step 2 using calculator to get the unknown currents.
- Step 4: Find the required branch current and power if required.

Problems

1. Consider a current of 0.55A is flowing through the resistance of 10Ω. Let us determine the potential difference between its ends.

Solution:

Given:

Current, I = 0.55A

Resistance, R = 10Ω

Formula to be used:

$$V = IR$$
$$= 0.55 \times 10$$
$$= 5.5\ V.$$

2. Consider a supply voltage of 200V is applied to a 100 Ω resistor. Let us determine the current flowing through it.

Solution:

Given:

Voltage, V = 200V

Resistance, R = 100Ω

Formula to be used:

$$\text{Current, } I = V / R$$
$$= 200 / 100$$
$$= 2A.$$

3. Let us calculate the resistance of the conductor if a current of 2A flows through it when the potential difference across its ends is 12V.

Solution:

Given:

Current, I = 2A

Potential difference, V = 12.

Formula to be used:

$$\text{Resistance, } R = V / I$$
$$= 12 / 2$$
$$= 6\,\text{ohm.}$$

4. Let us determine the current flowing in the 40Ω Resistor, R_3.

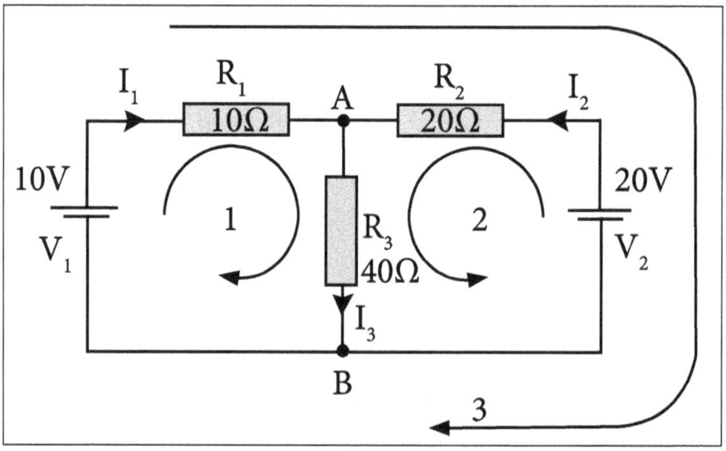

Solution:

Given:

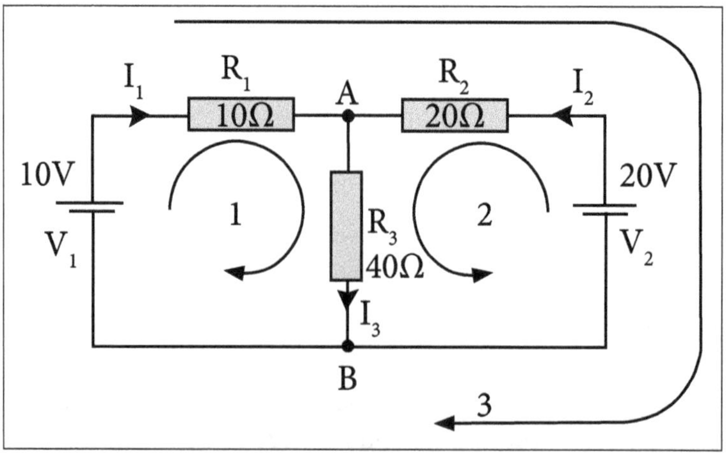

The circuit has three branches, two nodes (A and B) and two independent loops.

Formula to be used:

Using Kirchhoff's Current Law, KCL, the equations are given as:

At node A: $I_1 + I_2 = I_3$

At node B: $I_3 = I_1 + I_2$

Using Kirchhoff's Voltage Law, KVL, the equations are given as:

Loop 1 is given as, $10 = R_1 \times I_1 + R_3 \times I_3 = 10I_1 + 40I_3$ ...(1)

Loop 2 is given as, $20 = R_2 \times I_2 + R_3 \times I_3 = 20I_2 + 40I_3$ ...(2)

Loop 3 is given as, $10 - 20 = 10I_1 - 20I_2$ ...(3)

As, I_3 is the sum of $I_1 + I_2$

We can rewrite the equations as:

Equation (1): $10 = 10I_1 + 40(I_1 + I_2) = 50I_1 + 40I_2$

Equation (2): $20 = 20I_1 + 40(I_1 + I_2) = 40I_1 + 60I_2$

We now have two simultaneous equations that can be reduced to give us the value of both I_1 and I_2.

Substitution of I_1 in terms of I_2 gives us the value of I_1 as -0.143 Amps.

Substitution of I_2 in terms of I_1 gives us the value of I_2 as +0.429 Amps.

As,

$$I_3 = I_1 + I_2$$

The current flowing in resistor R_3 is given as:

$$-0.143 + 0.429 = 0.286 \text{ Amps.}$$

The voltage across the resistor R_3 is given as:

$$0.286 \times 40 = 11.44 \text{ volts.}$$

5. Let us determine the current through a 20Ω resistance and current through a 40Ω resistance.

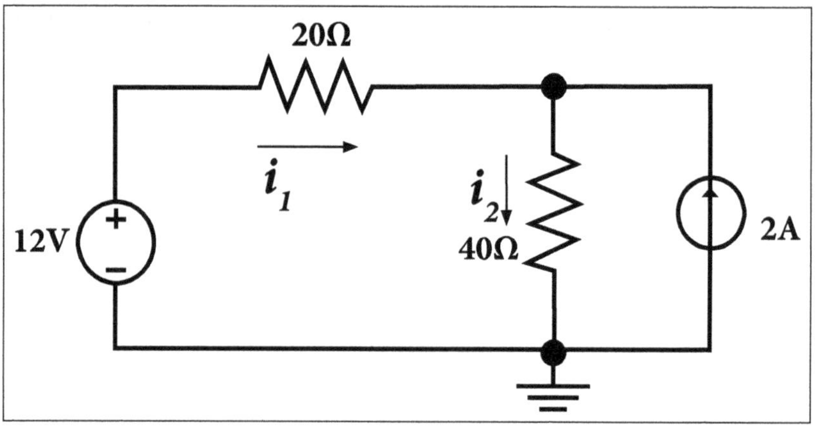

Solution:

Given:

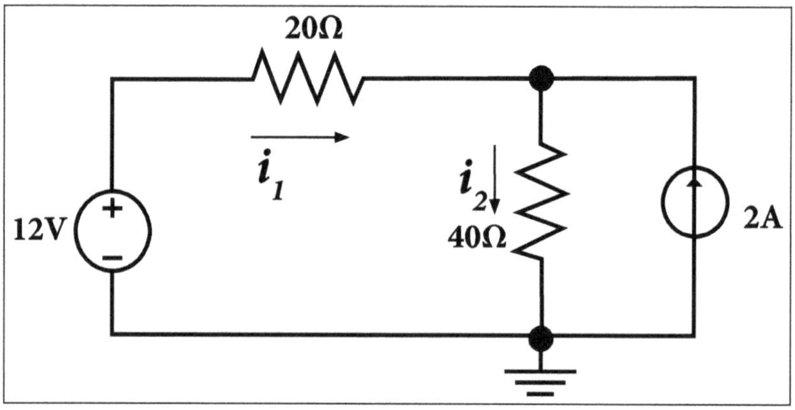

Write, KCL at node x:

$$-i_1 + i_2 - 2A = 0$$

Write, v_x in the circuit using Ohm's Law:

$$i_1 = \frac{12V - v_x}{20\Omega}, \quad i_2 = \frac{v_x}{40\Omega}$$

Apply last two equations into KCL at node x:

$$-i_1 + i_2 - 2A = -\frac{12V - v_x}{20\Omega} + \frac{v_x}{40\Omega} - 2A = 0$$

$$-0.6 + 0.05v_x + 0.025v_x - 2 = 0$$

$$v_x = 34.67\,V$$

The current through a 20Ω resistance:

$$i_1 = \frac{12V - 34.67}{20\Omega} = -1.133\,A$$

The current through a 40Ω resistance:

$$i_2 = \frac{v_x}{40\Omega} = \frac{34.67}{40\Omega} = 0.866\,A$$

6. Let us find the current in a circuit using Kirchhoff's voltage law.

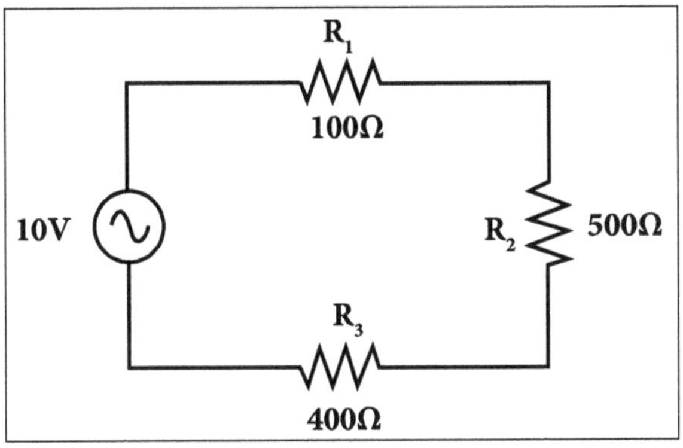

Solution:

Given:

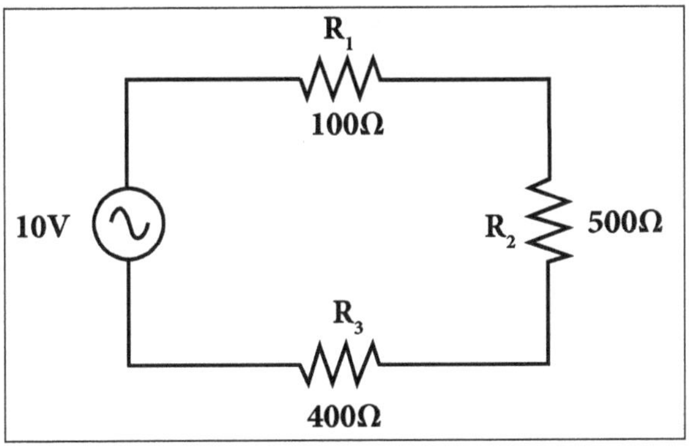

10 = 100(I) + 500(I) + 400(I)

10 = 1000(I)

I = 10/1000

I = 0.01A

7. Let us determine the current i and voltage v over each resistor.

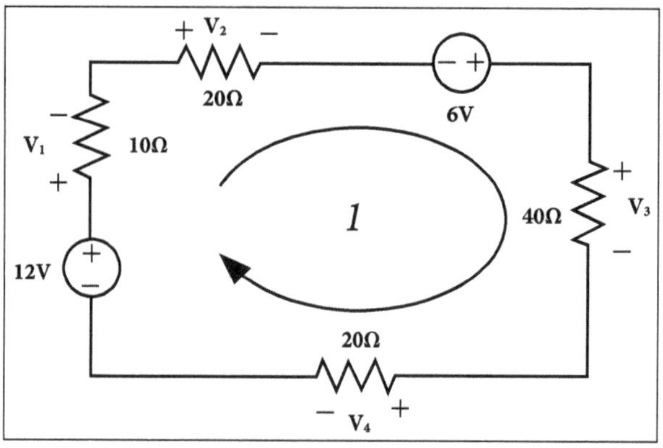

Solution:

Given:

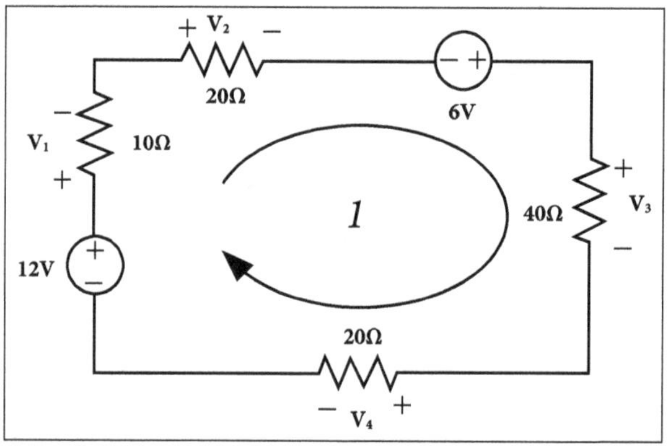

KVL equations for voltages:

$$v_1 + v_2 + v_3 + v_4 = 18$$

Using Ohm's Law:

$$v_1 = 10\Omega,\ v_2 = 20\Omega,\ v_3 = 40\Omega,\ v_4 = 20\Omega$$

Substituting into KVL equation, we get:

$$10i + 20i + 40i + 20i = 18$$

$$90i = 18$$

$$i = \frac{18}{90} = 0.2\,A$$

$$v_1 = R_1 i = 10(0.2) = 2V$$

$$v_2 = R_2 i = 20(0.2) = 4V$$

$$v_3 = R_3 i = 40(0.2) = 8V$$

$$v_4 = R_4 i = 20(0.2) = 4V$$

8. Let us determine v_1 and v_2 in the following circuit.

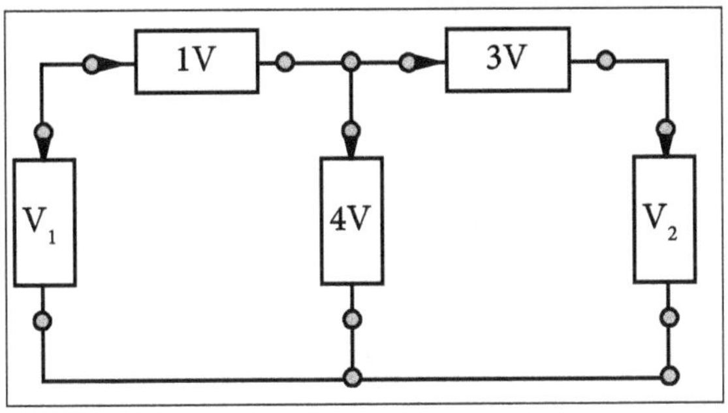

Solution:

Given:

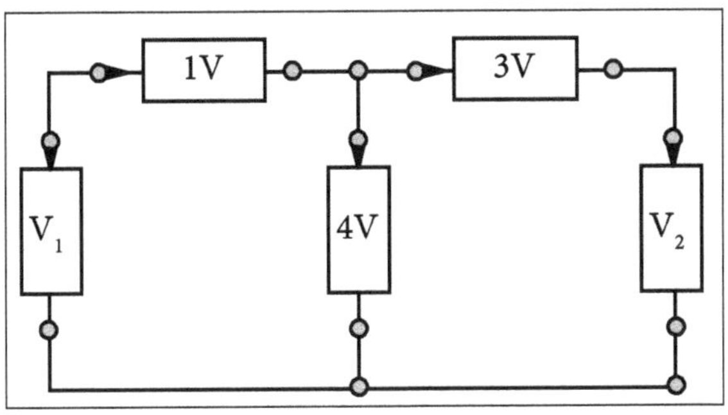

Loop 1:

$$-V_1 + 1 + 4 = 0$$

$$V_1 = 5V$$

Loop 2:

$$-4 + 3 + V_2 = 0$$

$$V_2 = 1V$$

9. Let us determine V_1, V_2 and V_3 where the arrows are signifying the positive position of the box and the negative is at the end of the box.

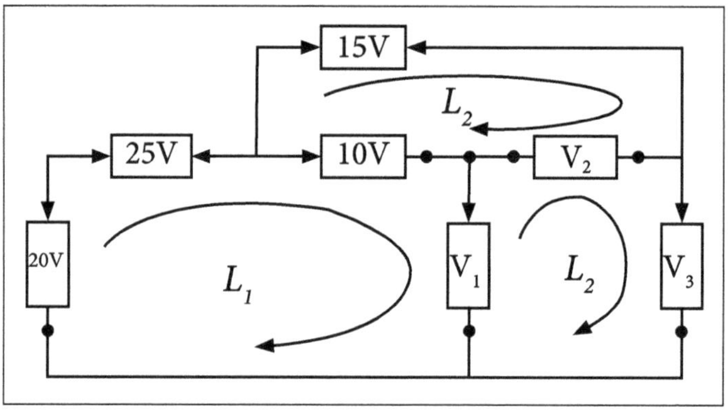

Solution:

Given:

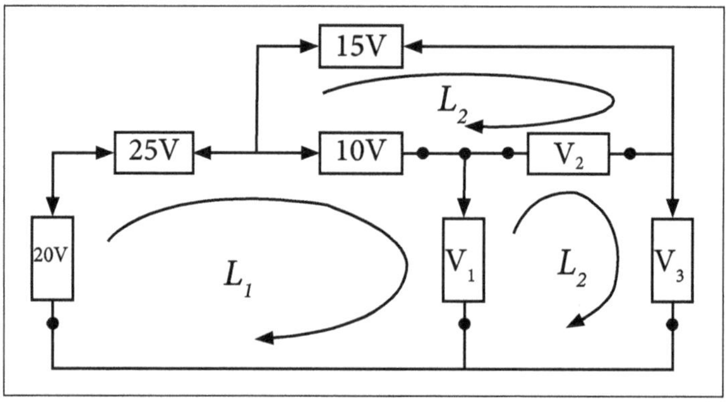

Loop 1:

$$-20-25+10+V_1=0$$

$$V_1=35\,V$$

Loop 2:

$$-10+15-V_2=0$$

$$V_2=5\,V$$

Loop 3:

$$-V_1+V_2+V_3=0$$

$$-35+5+V_3=0$$

$$V_3=30\,V$$

10. Let us determine V_1, V_2, V_3 and V_4 from the figure in which the arrows are signifying the positive position of the box and the negative is at the end of the box.

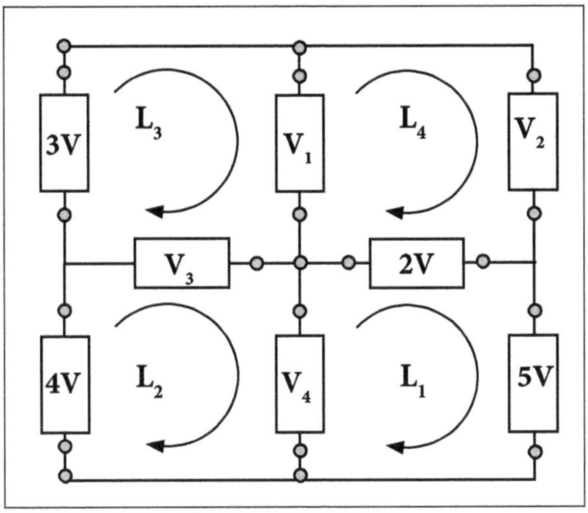

Solution:

Given:

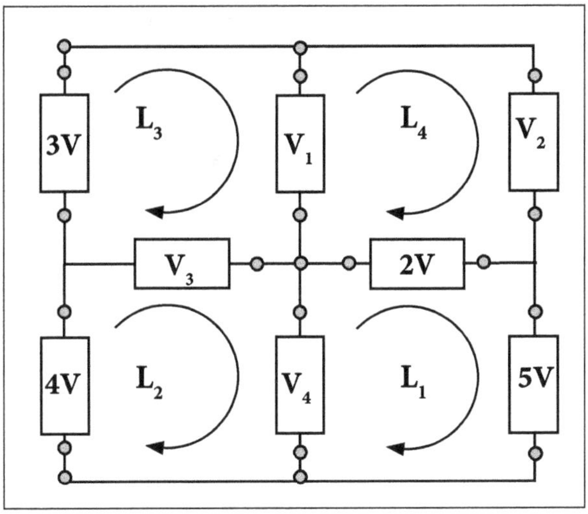

Loop 1:

$$-V_4 + 2 + 5 = 0$$

$$V_4 = 7V$$

Loop 2:

$$4 + V_3 + V_4 = 0$$

$$V_3 = -4-7$$

$$V_3 = -11\,V$$

Loop 3:

$$-3+V_1-V_3=0$$

$$V_1 = V_3+3$$

$$V_1 = -8\,V$$

Loop 4:

$$-V_1-V_2-2=0$$

$$V_2 = -V_1-2$$

$$V_2 = 6\,V$$

11. Let us determine V_1, V_2 and V_3 in the following circuit.

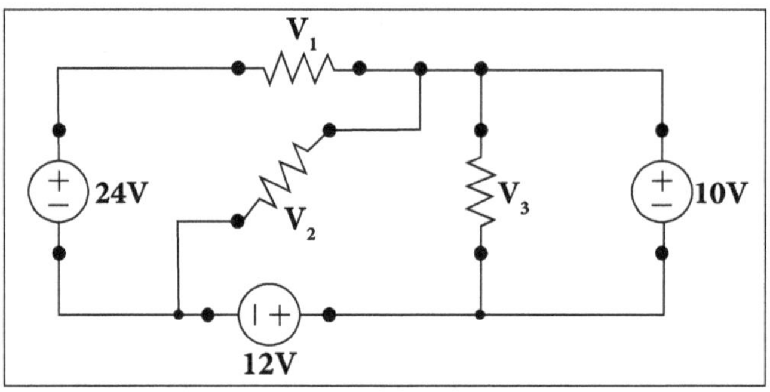

Solution:

Given:

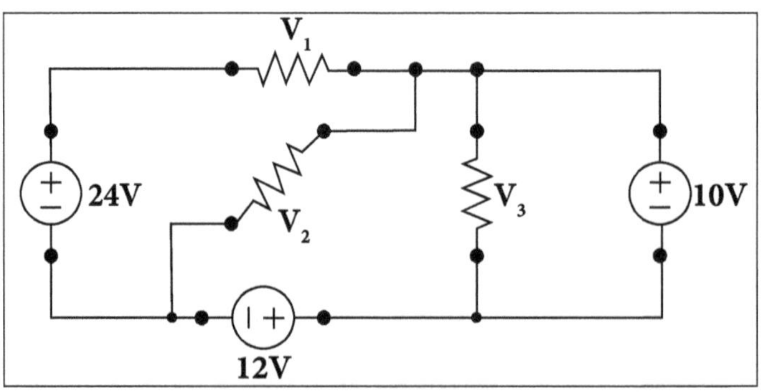

Outer loop:

$$-24+V1+10+12=0$$

$$V_1=2V$$

V_2 and 10V source loop:

$$V_2+10+12=0$$

$$V_2=-22V$$

V_3 and 10V source loop:

$$-V_3+10=0$$

$$V_3=10V$$

12. Let us determine I_1, I_2, I_3 in the following circuit.

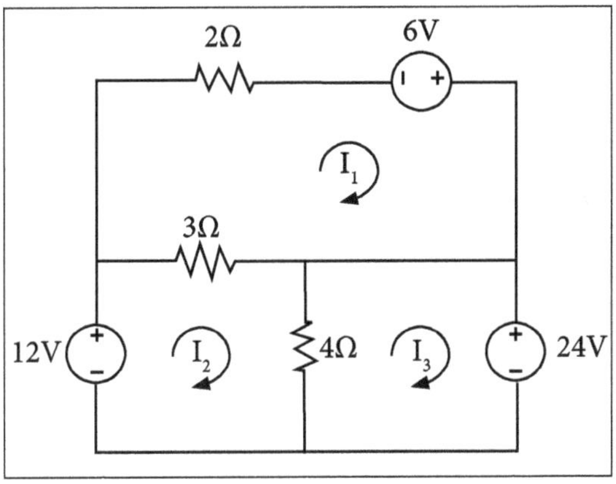

Solution:

Given:

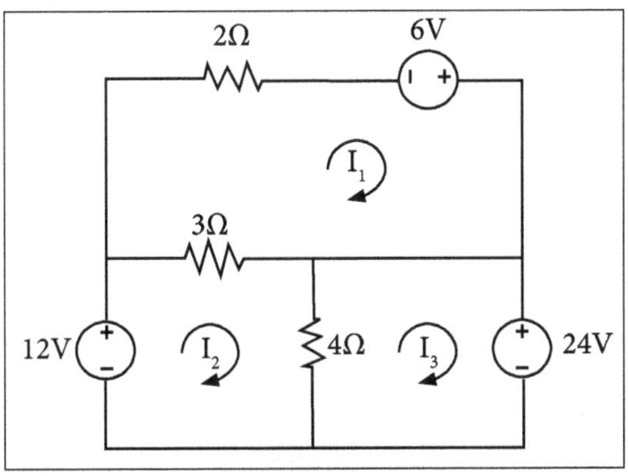

Loop 1:

$$-6 = -2\,I_1 + 3(I_2 - I_1)$$

Loop 2:

$$12 = 3(I_2 - I_1) + 4(I_2 - I_3)$$

Loop 3:

$$24 = -4(I_3 - I_2)$$

From Loop 3:

$$(I_2 - I_3) = 6$$

Substitute in loop 2, the results is:

$$12 = 3(I_2 - I_1) + 4(6)$$

$$I_2 - I_1 = -4$$

Substitute in loop 1, the results is:

$$-6 = -2I_1 + 3(-4)$$

$$I_1 = -3A$$

Substitute in loop 2, the results is:

$$I_2 - (-3A) = -4A$$

$$I_2 = -7A$$

Substitute in loop 3, the results is:

$$24 = -4(I_3 + 7)$$

$$-6 = I_3 + 7$$

$$I_3 = -13A$$

Superposition Theorem is one of the electrical network analysis theorem which helps us to solve the linear circuit with more than one current or voltage source easily.

In a linear circuit with several sources, the current and voltage responses in any branch is the algebraic sum of the voltage and current responses due to each of the source acting independently with all the other sources replaced by their internal impedance.

Suppose an electrical circuit having several branches or loads and also several source some being current source and some being voltage source. Then Superposition theorem suggests that:

If we find the branch responses on a branch due to only of those source by ignoring the effect of all the other sources or replacing all the other sources by their corresponding internal impedance and repeat the process for every source on the circuit, then the combined responses (Voltage drop and Current through it) on a branch due to all the sources combined is the algebraic sum of the responses on the branches due to each individual sources.

Steps to be followed to apply the superposition theorem:

- Step 1: Select any one of the energy source and replace all the other energy sources by their internal series resistances.
- Step 2: With only one energy source, let us calculate the current through the required branch using series parallel along with current division rule or else using mesh analysis.
- Step 3: Repeat the steps 1 and 2 for each source individually.
- Step 4: Algebraically add the currents obtained due to the individual source to obtain the combined effect of all the sources.

Problem

1. Let us determine the current I_{ab} flowing through the 3Ω resistor as shown in the below figure using the superposition theorem.

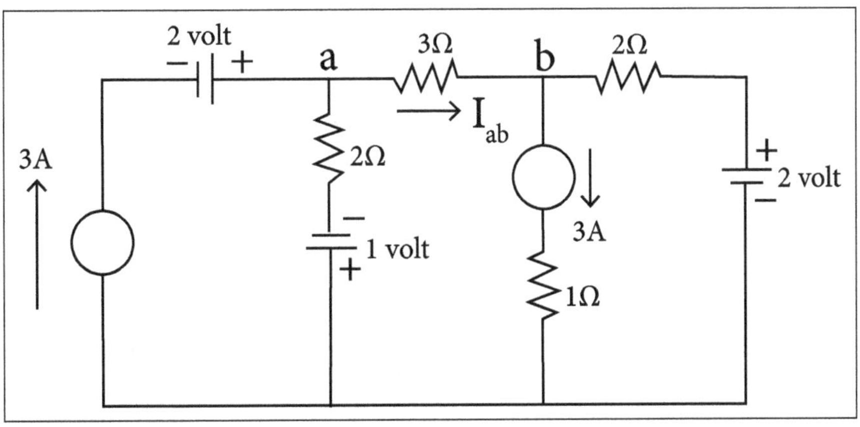

Solution:

Given:

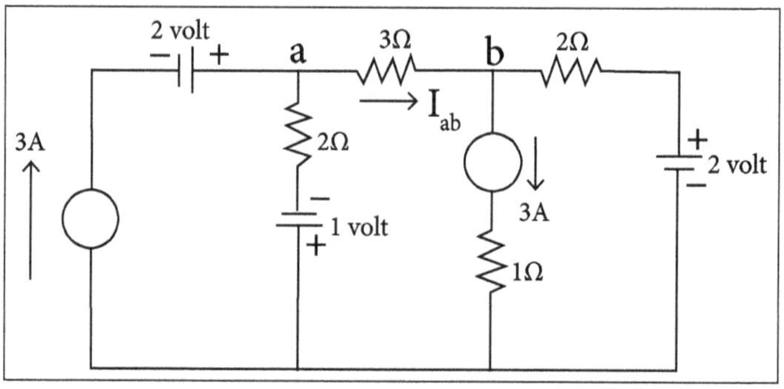

(a)

Assume that the current source 3A is acting alone in the circuit and the internal resistances replace the other sources.

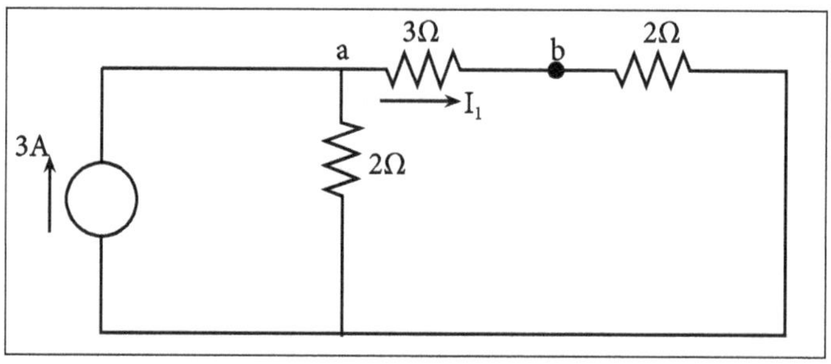

(b)

The current flowing through 3Ω resistors can be obtained from figure (b) and it is given by:

$$I_{1(\text{due to 3A current source})} = 3\times\frac{2}{7} = \frac{6}{7}A(a\,to\,b) \qquad ...(1)$$

Current flowing through 3Ω resistor due to 2V source can be obtained from figure (c).

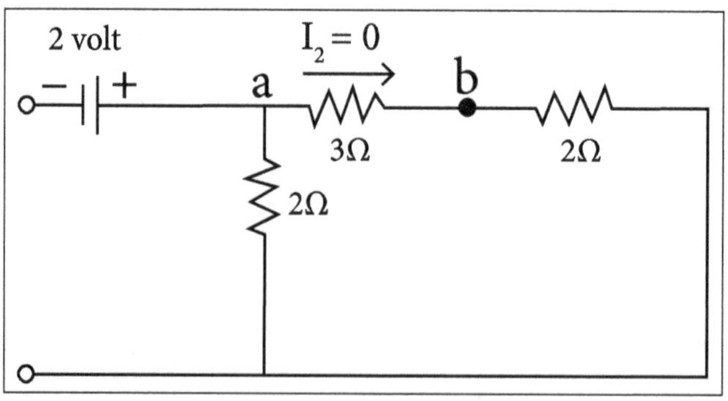

(c)

It is seen that no current is flowing.

$$I_{2(\text{due to 2V voltage source})} = 0A(\text{a to b})$$

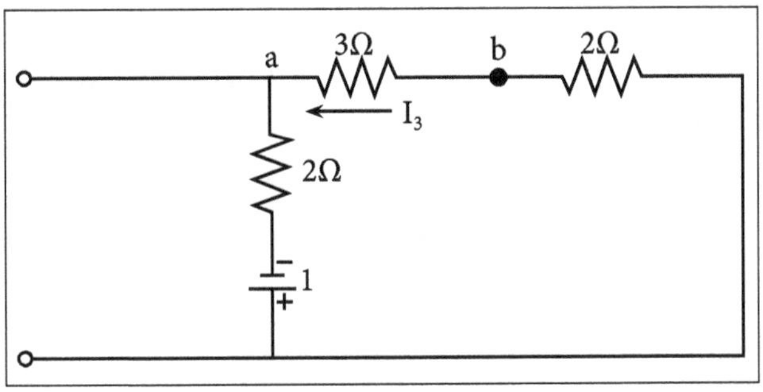

(d)

Current through 3Ω resistor due to 1V voltage source only is given by:

$$I_{3(\text{due to 1V voltage source})} = 1/7\,A(\text{b to a})$$

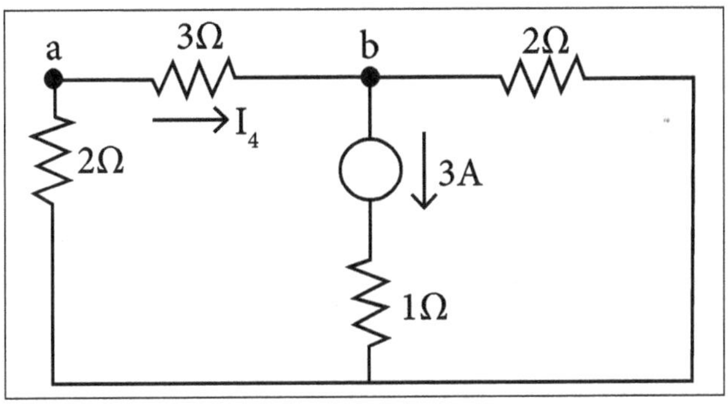

(e)

From figure (e), Current through 3Ω resistor due to 3A current source only is obtained by:

$$I_{4(\text{due to 3A current source})} = 3\times\frac{2}{7} = \frac{6}{7}A(\text{a to b})$$

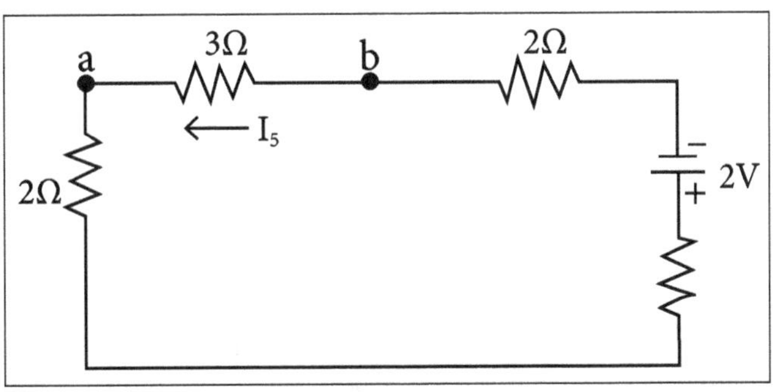

(f)

From figure (f), Current through 3Ω resistor due to 2V voltage source only is given by:

$$I_{5(\text{due to 2V voltage source})} = \frac{2}{7}A(b\,to\,a) \quad ...(2)$$

Resultant current I_{ab} flowing through 3Ω resistor due to the combination of all sources is obtained by the following expression:

$$I_{ab} = I_{1(\text{dueto 3A current source})} + I_{2(\text{due to 2V current source})} + I_{3(\text{due to 1V current source})} + I_{4(\text{dueto 3A current source})} + I_{5(\text{due to 3A current source})}$$

$$= \frac{6}{7} + 0 - \frac{1}{7} + \frac{6}{7} - \frac{2}{7} = \frac{9}{7} = 1.285(a\,to\,b)$$

1.2.2 Source Equivalence and Conversion

Source Conversion

Source transformation is defined as converting the sources for easy analysis of the circuit. In mesh analysis, it is easier if the circuit has voltage sources. In nodal analysis, it is easier if the circuit has current sources.

Voltage source to current source transformation: If voltage source is converted to current source, then the current source $I_S = V_S / R_S$ with parallel resistance equal to R_s.

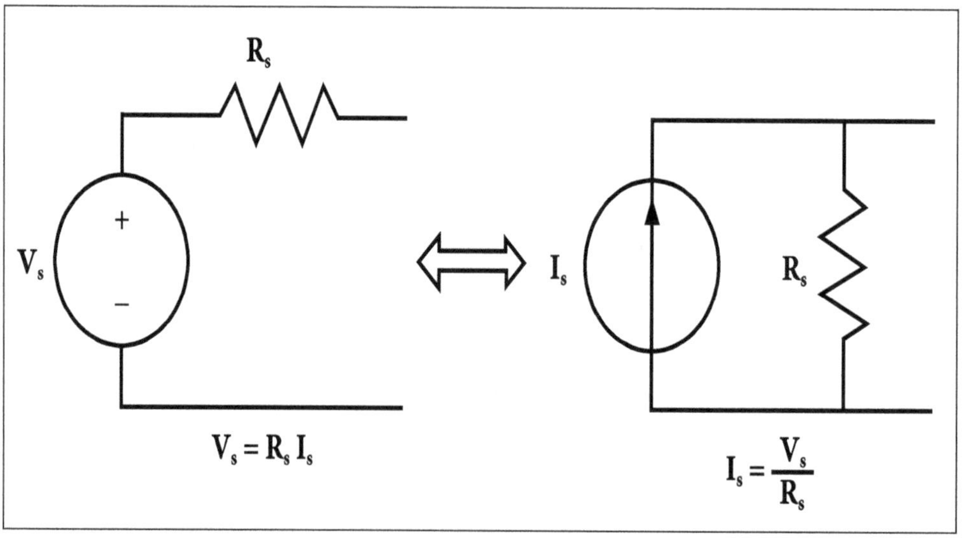

Voltage Source to Current Source.

Current source to voltage source transformation: If current source is converted to voltage source, then the voltage source $V = IR_{sh}$ with series resistance equal to R_{sh}.

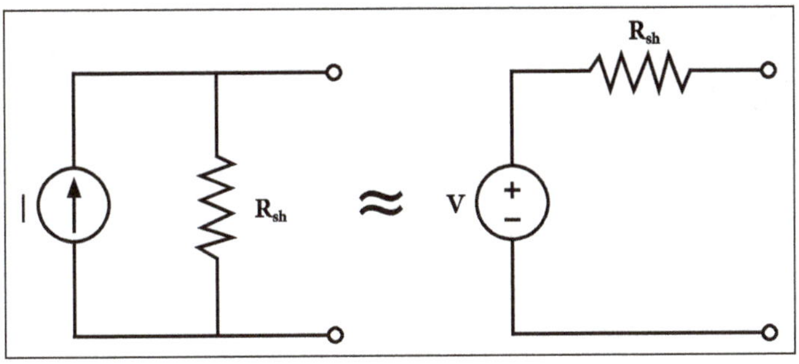

Current Source to Voltage Source.

Problem

Let us determine the current through 16Ω resistor for the given diagram.

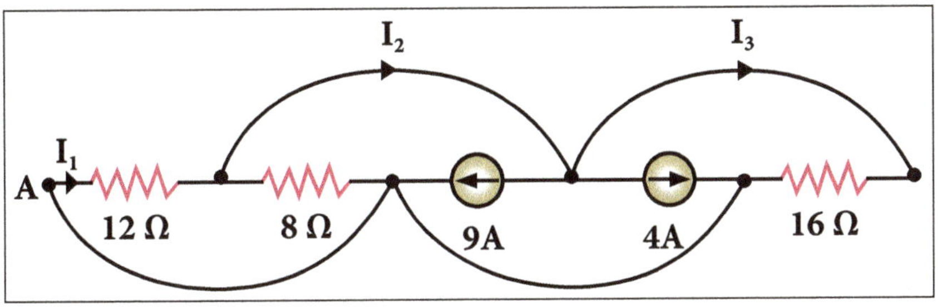

Solution:

Given:

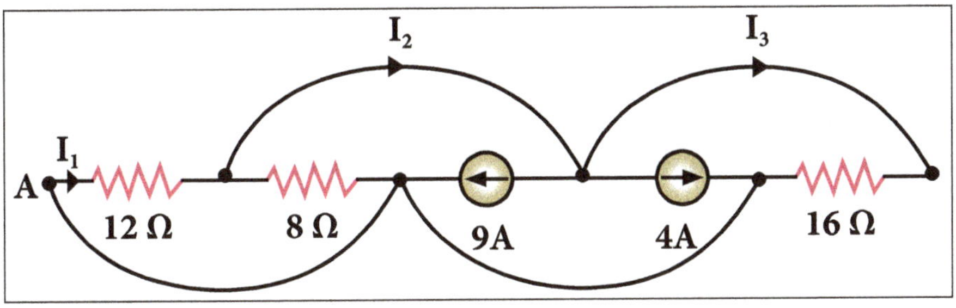

Assigning currents to each branch using KCL:

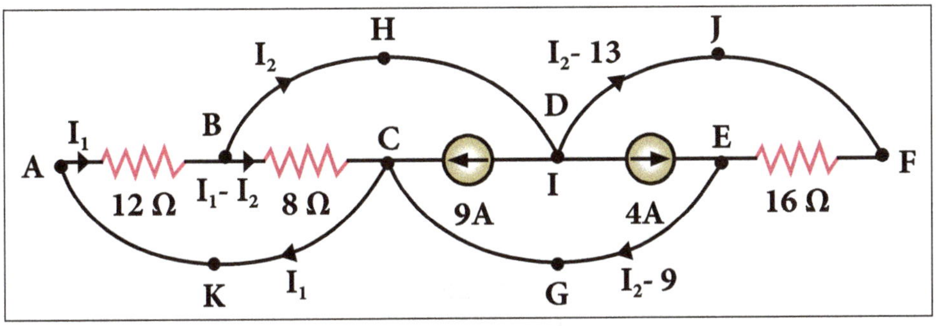

Applying KVL to ABCKA:

$$-12\,I_1 - 8(I_1 - I_2) = 0$$

$$-\,20\,I_1 + 8\,I_2 = 0 \qquad ...(1)$$

Applying KVL to ABHDJFEGCKA:

$$-\,12\,I_1 - 16(I_2 - 1_3) = 0$$

$$-12\,I_1 - 16I_2 = -208 \qquad ...(2)$$

Solving equations (1) and (2), we get:

$$I_1 = 4\,A(\rightarrow)$$

$$I_2 = 10A(\curvearrowright)$$

$$I_{16\Omega} = I_2 - 1_3$$

$$= 10 - 13 = -3A$$

$$I_{16\Omega} = -3A$$

1.2.3 Thevenin's Theorem

In circuit theory, Thevenin's theorem for linear electrical networks states that any combination of voltage sources, current sources and resistors with two terminals is electrically equivalent to the single voltage source V and the single series resistor R. For a single frequency AC system, the theorem can also be applied to general impedances, not just for resistors.

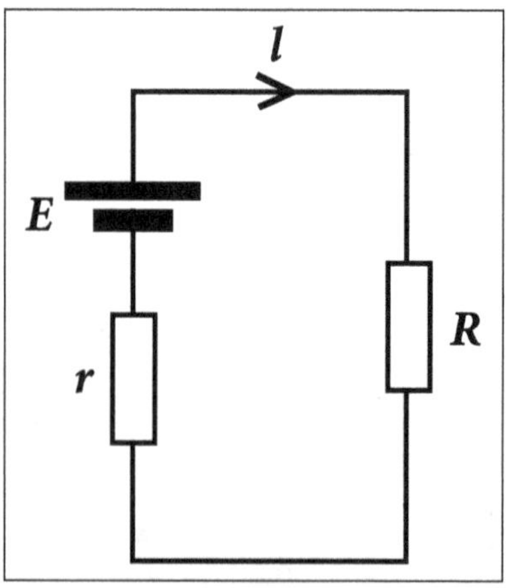

Thevenin's circuit.

'To determine the current in any branch of the active network.'

The procedure adopted when using the Thevenin's theorem is summarized as follows:

- Remove the resistance R from that branch.
- Determine the open-circuit voltage, E, across the break.
- Remove each of the source of e.m.f. and replace them by their internal resistances and then determine the resistance r, 'looking-in' at the break.
- Determine the value of the current from the equivalent circuit shown in the figure:

  i.e., $I = ER + r$

Problem

1. Let us determine the current in the 0.8 Ω resistor using Thevenin's theorem for the below circuit.

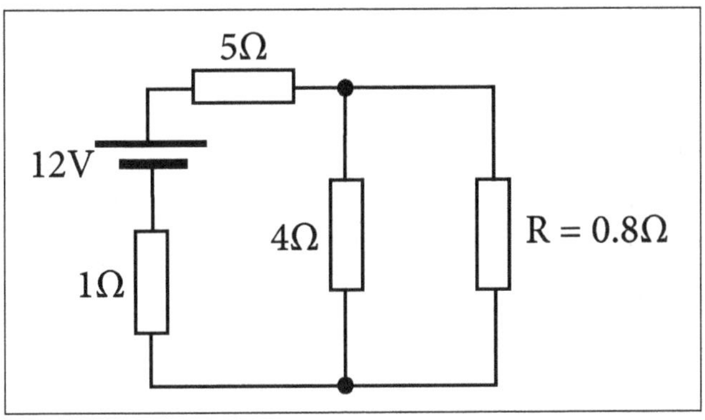

Solution:

Given:

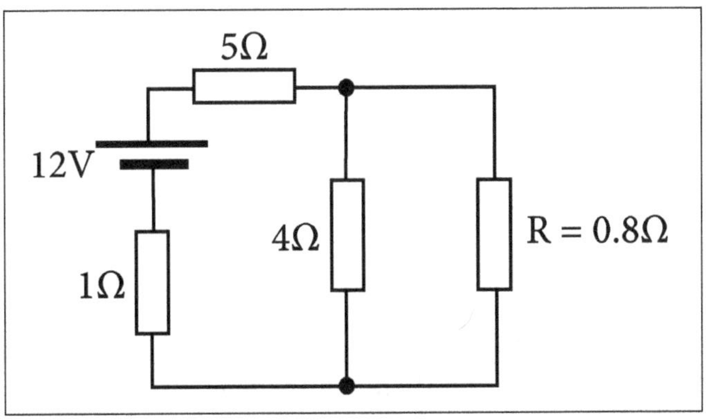

By following the above procedure:

(i) The 0.8Ω resistor is removed from the circuit as shown in figure:

Current, $I_1 = 12/(1+5+4)$

$= 12/10$

$= 1.2A$

(ii) P.d. across 4Ω resistor = $4I_1 = (4)(1.2) = 4.8V$

Hence, p.d. across AB, i.e., the open-circuit voltage across AB:

E = 4.8V.

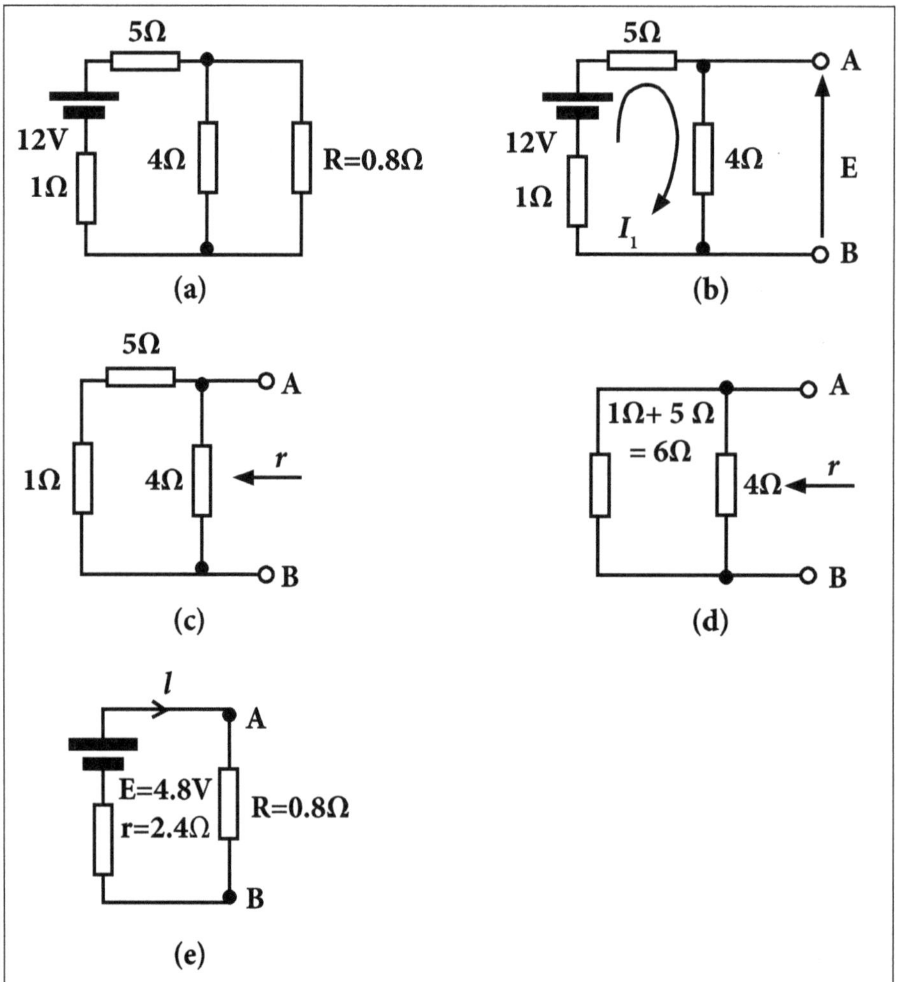

(iii) Removing the source of e.m.f. gives the circuit shown in figure (c). The equivalent circuit of figure (c) is shown in figure (d), from which resistance r = 4×6/4+6= 24/10 =2.4 Ω.

(iv) The equivalent Thevenin's circuit is shown in figure (e).

From the circuit (e):

$$\text{Current } I = E/(r + R)$$

$$= 4.8/(2.4+0.8)$$

$$= 4.8/3.2$$

$I = 1.5A$ = Current in the 0.8Ω resistor.

1.2.4 Norton's Theorem

Norton's theorem states the following:

Any two-terminal linear bilateral DC network can be replaced by an equivalent circuit consisting of a current and a parallel resistor.

Preliminary Steps

The steps leading to the proper values of I_N and R_N are as follows:

- Remove the portion of the network across which the Norton equivalent circuit is found.
- Mark the terminals of the remaining two-terminal network.
- Finding R_N:

Calculate the value of R_N by first setting all sources to zero and then finding the resultant resistance between the two marked terminals. Since $R_N = R_{Th}$, the procedure and value obtained using the approach described for Thevenin's theorem will determine the proper value of R_N.

- Finding I_N:

Calculate the value of I_N by first returning all the sources to their original position and then finding the short-circuit current between the marked terminals. It is the same amount of current that can be measured by an ammeter which is placed between the marked terminals.

- Finally, draw the Norton equivalent circuit with the portion of the circuit previously removed or replaced between the terminals of the equivalent circuit.

Problem

1. Let us determine the current I flowing in the 4Ω resistance shown in figure (a) by using Norton's theorem.

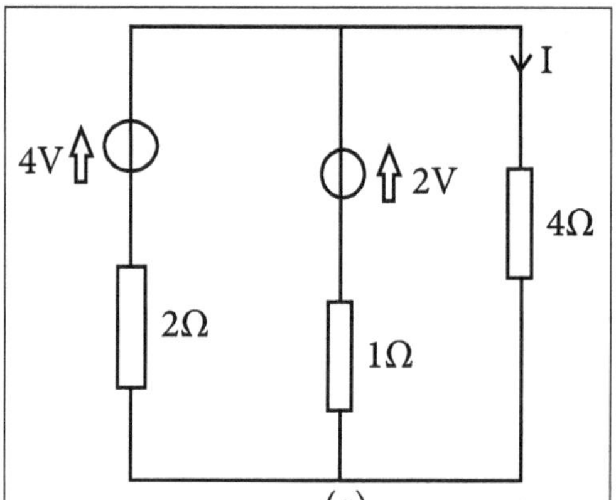

(a)

Solution:

Given:

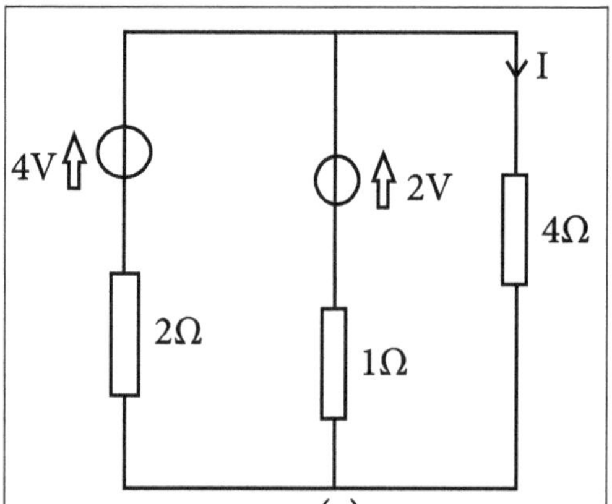

(a)

The 4Ω branches is short-circuited as shown in figure (b).

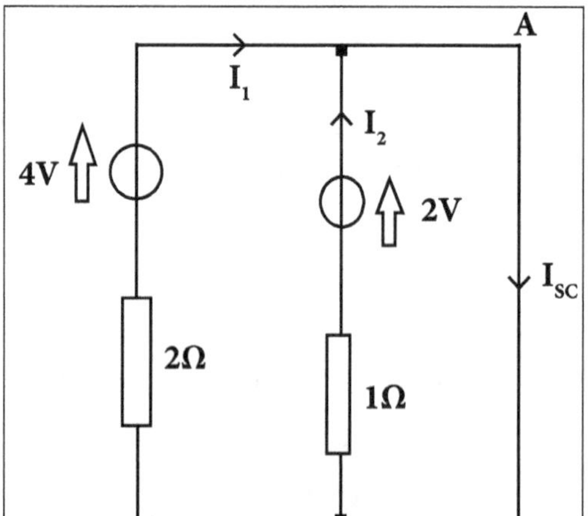

(b)

From figure (b):

$$I_{SC} = I_1 + I_2 = 4A.$$

If the sources of e.m.f. are removed, the resistance 'looking-in' at a break made between A and B is given by:

$$r = 2\times1/2+1 = 2/3\Omega$$

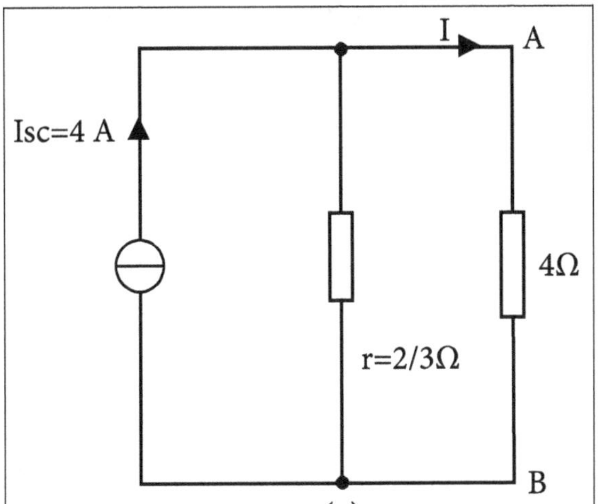

(c)

From the Norton equivalent network shown in figure (c), the current in the 4Ω resistance is given by:

$$I = \left[(2/3)/((2/3)+4)\right](4) = 0.571A.$$

1.3 Nodal Analysis and Mesh Analysis

Mesh Current Method

A 'mesh' is any closed path in a given circuit that does not have any element (or branch) inside it. A mesh has the properties such as:

- Every node in the closed path is exactly formed with two branches.
- No other branches are enclosed by the closed path.

Mesh Analysis

This is an alternative structured approach for solving the circuit and is based on calculating mesh currents. A similar approach to the node situation is used. A set of equations (based on KVL for each mesh) is formed and the equations are solved for unknown values. As many equations are needed as unknown, mesh currents exist.

Steps for Analysing the Mesh Current:

- Step 1: Identify the mesh currents.
- Step 2: Determine which mesh currents are known.
- Step 3: Write equation for each mesh using KVL and that includes the mesh currents.
- Step 4: Solve the equations.

Node Voltage Method

In the node voltage method, we identify all the nodes on the circuit. Choosing one of them as the reference voltage (i.e., zero potential) and subsequently assign other node voltages (unknown) with respect to the reference voltage.

If the circuit has 'n' nodes, there are 'n-1' node voltages are unknown. At each of these 'n-1' nodes, we can apply KCL equation. The unknown node voltages become the independent variables of the problem and the solution of node voltages can be obtained by solving a set of simultaneous equations.

Nodal Analysis

Nodal analysis involves looking at a circuit and determining all the node voltages in the circuit. The voltage at any given node of circuit is the voltage drop between that node and a reference node (usually ground). Firstly, all the nodes in the circuited are counted and identified. Secondly, nodes at which the voltage is already known are listed.

A set of equations based on the node voltages are formed and these equations are solved for unknown quantities. The set of equations are formed using the KCL equations at each node. The set of simultaneous equations that is produced is then solved. Then, branch currents can be found once the node voltages are known. This can be reduced to a series of steps:

- Step 1: Identify the nodes.
- Step 2: Choose a reference node.
- Step 3: Identify which node voltages are known if any.
- Step 4: Identify the branch currents.

- Step 5: Use KCL to write an equation for each unknown node voltage.
- Step 6: Solve the equations.

Problems

1. For the circuit shown in the below figure, let us determine V_x using the mesh current method.

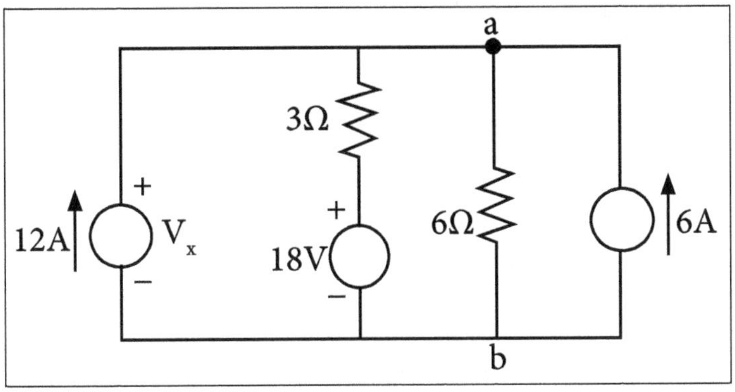

Solution:

Given:

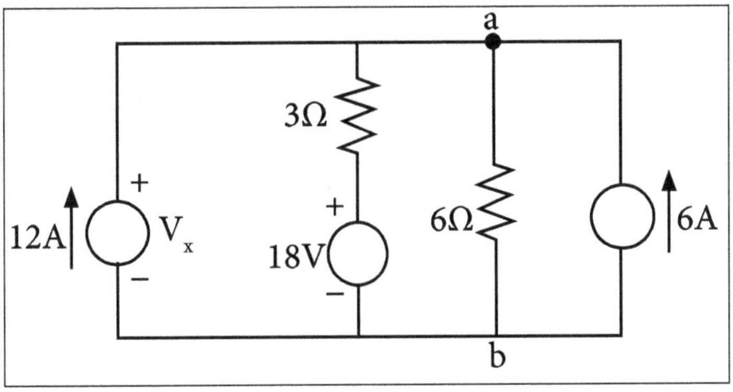

Convert the extreme right current source (6A) into a voltage source. Note that the current source magnitude is 6A and its internal resistance is 6Ω.

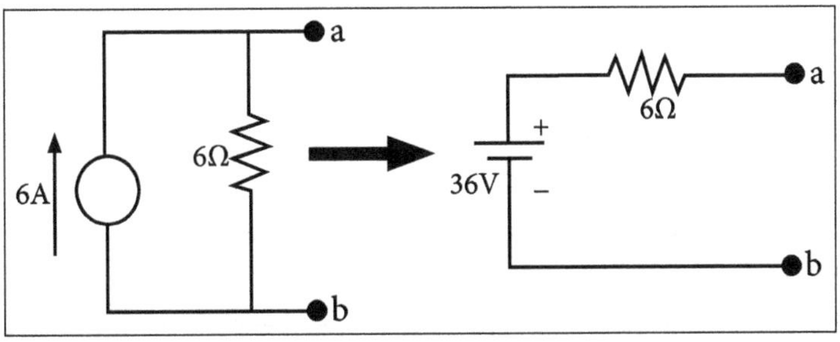

The given circuit is redrawn and shown in the below figure:

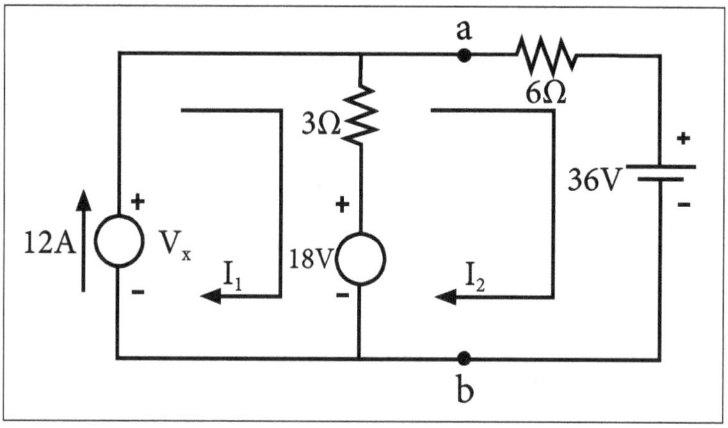

Loop 1:

By using KVL:

$$V_x - (I_1 - I_2) \times 3 - 18 = 0 \Rightarrow V_x + 3\,I_2 = 54 \quad ...(1)$$

Where, I_1 = 12A

Loop 2:

By using KVL:

$$18 - (I_2 - I_1) \times 3 - I_2 \times 6 - 36 = 0 \Rightarrow 9\,I_2 = 18 \Rightarrow I_2 = 2A \quad ...(2)$$

Substituting equation (2) in equation (1), we get:

$$V_x = 48 \text{ volts}$$

2. Let us determine the current through branch a-b using mesh analysis.

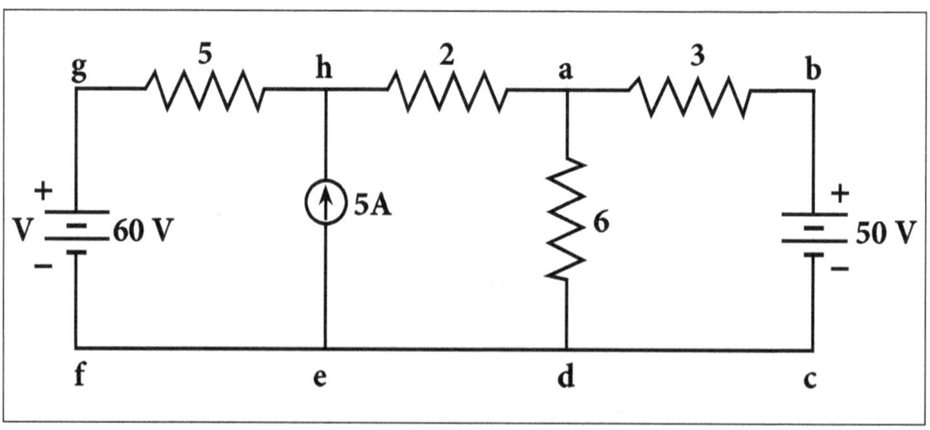

Solution:

Given:

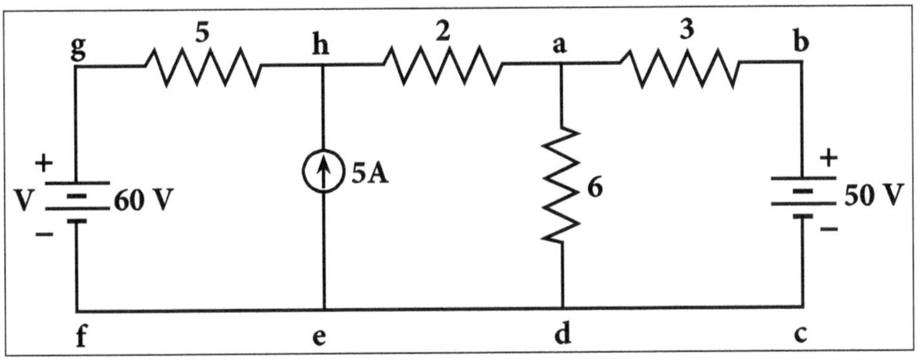

Consider the loops:

Loop HADEH:

$$5I_1 + 2I_2 + 6(I_2 - I_3) = 60$$

$$5I_1 + 8I_2 - 6I_3 = 60 \quad ...(1)$$

Loop ABCDA:

$$3I_3 + 6(I_3 - I_2) = -50$$

$$3I_3 + 6I_3 - 6I_2 = -50$$

$$9I_3 - 6I_2 = -50 \quad ...(2)$$

$$I_2 - I_1 = 5A \quad ...(3)$$

From (1), (2) & (3),

$$D = \begin{vmatrix} -1 & 1 & 0 \\ 5 & 8 & -6 \\ 0 & -6 & 9 \end{vmatrix}$$

$$= -1(72-36) - 1(45)$$

D = -81.

$$D_3 = \begin{vmatrix} -1 & 1 & 5 \\ 5 & 8 & 60 \\ 0 & -6 & -50 \end{vmatrix}$$

$$= -1(-400+360)-(-250) + 5(-30)$$

$$= 40+250-150$$

$$D_3 = 140.$$

$$I_3 = D_3 / D$$

$$= 140/-81$$

$$= -1.7283$$

The current through the branch ab is 1.7283A.

3. Let us compute V_1 and V_2 in the circuit shown below, using nodal analysis.

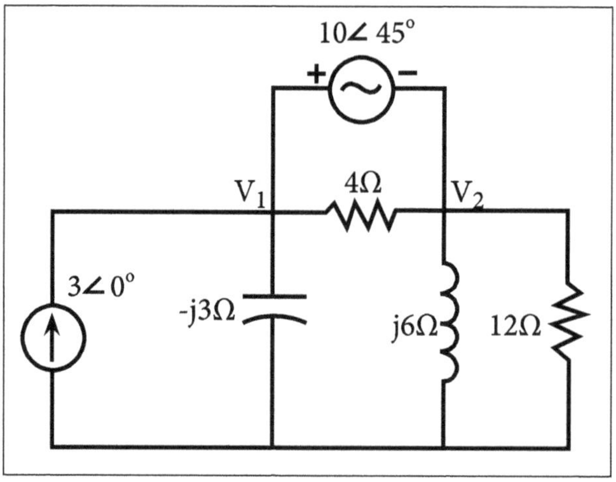

Solution:

Given:

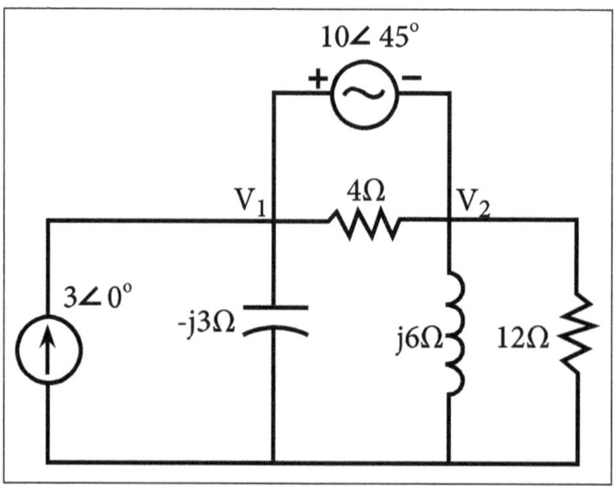

Step 1: Given voltage difference between V_1 & V_2 is:

$$V_1 - V_2 = 10\angle 45^\circ$$

$$V_1 - V_2 = 7.07 + j7.07 \quad ...(1)$$

Step 2: By applying KCL at nodes V_1 & V_2:

$$\frac{V_1}{-j3} + \frac{V_2}{j6} + \frac{V_2}{12} = 3\angle 0^\circ$$

$$V_1(j0.33) + V_2\left(\frac{1}{j6} + \frac{1}{12}\right) = 3\angle 0^\circ$$

$$V_1(j0.33) + V_2(0.0833 - 0.1666j) = 3 + j0 \quad ...(2)$$

Step 3: From equation (1) and (2):

$$\begin{bmatrix} 1 & -1 \\ j0.333 & 0.0833 - j0.1666 \end{bmatrix}\begin{bmatrix} V_1 \\ V_2 \end{bmatrix} = \begin{bmatrix} 7.07 + j7.07 \\ 3 \end{bmatrix}$$

Step 4: To determine V_1 and V_2:

$$V_1 = \frac{\Delta_1}{\Delta} = \frac{\begin{bmatrix} 7.07 + j7.07 & -1 \\ 3 & 0.0833 - j0.1666 \end{bmatrix}}{\begin{bmatrix} 1 & -1 \\ j0.333 & 0.0833 - j0.1666 \end{bmatrix}}$$

$$= \frac{(0.0833 - j0.1666)*(7.07 + j7.07) + 3}{(0.0833 - j0.1666) + j0.333}$$

$$= \frac{4.7665 - j0.5886}{0.0833 + j0.1664}$$

$$= 8.636 - j24.322$$

$$V_1 = \frac{\Delta_2}{\Delta} = \frac{\begin{bmatrix} 1 & 7.07 + j7.07 \\ j0.333 & 3 \end{bmatrix}}{\begin{bmatrix} 1 & -1 \\ j0.333 & 0.0833 - j0.1666 \end{bmatrix}}$$

$$= \frac{3-(7.07+j7.07)*(j0.33)}{(0.0833-j0.1666)+j0.333}$$

$$= \frac{5.354-j2.354}{0.0833+j0.1666}$$

$$= 1.567 - j31.390$$

The node voltages V_1 and V_2 are in polar form as:

$$V_1 = 8.636 - j24.322 \text{ (or) } 25.80\angle -70.45^\circ$$

$$V_2 = 1.567 - j31.390 \text{ (or) } 31.429\angle -87.14^\circ$$

4. Let us use mesh analysis to determine three mesh currents in the circuit shown below.

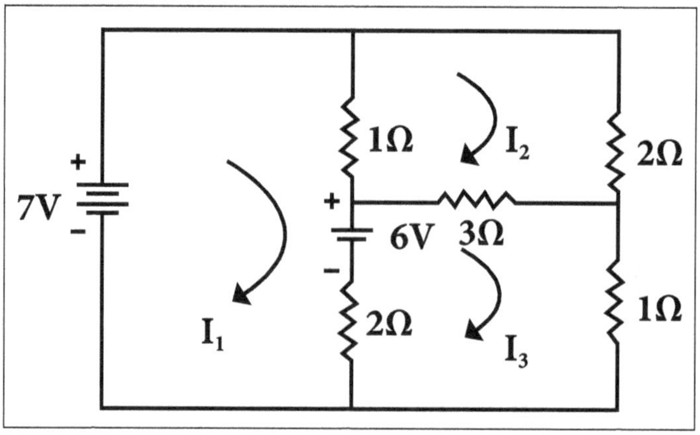

Solution:

Given:

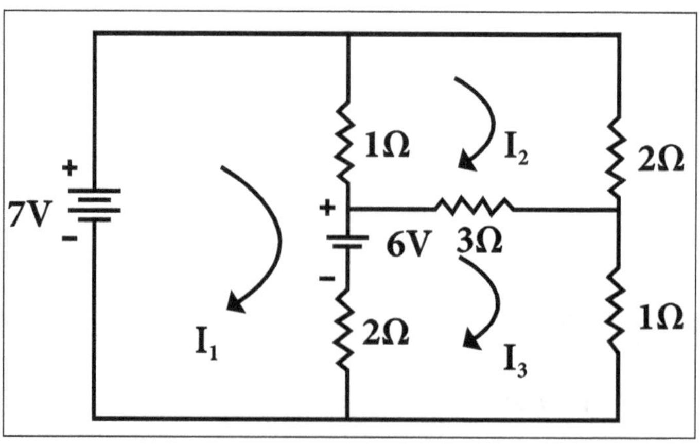

Mesh Loop 1:

$$-1(I_1-I_2)-6-2(I_1-I_3)+7=0$$

$$-3I_1-I_2-2I_3=1 \qquad ...(1)$$

Mesh Loop 2:

$$-2I_2-3(I_2-I_3)-1(I_2-I_1)=0$$

$$-I_1+6I_2-3I_3=0 \qquad ...(2)$$

Mesh Loop 3:

$$I_3-2(I_3-I_1)+6-3(I_3-I_2)=0$$

$$-2I_1-3I_2+6I_3=0 \qquad ...(3)$$

Let us form the matrix from (1), (2) and (3):

$$\begin{bmatrix} 3 & -1 & -2 \\ -1 & 6 & -3 \\ -2 & -3 & 6 \end{bmatrix}\begin{bmatrix} I_1 \\ I_2 \\ I_3 \end{bmatrix}=\begin{bmatrix} 1 \\ 0 \\ 6 \end{bmatrix}$$

By applying Cramer's rule:

$$\Delta=\begin{bmatrix} 3 & -1 & -2 \\ -1 & 6 & -3 \\ -2 & -3 & 6 \end{bmatrix}$$

$$=3(36-9)-(-1)(-6-6)-2(3+12)$$

$$=39.$$

To find I_1:

$$\Delta_1=\begin{bmatrix} 1 & -1 & -2 \\ 0 & 6 & -3 \\ 6 & -3 & 6 \end{bmatrix}=1(36-9)-(-1)(18)-2(-36)$$

$$=117$$

$$I_1=\frac{\Delta_1}{\Delta}=\frac{117}{39}\Rightarrow 3A.$$

To find I_2:

$$\Delta_2 = \begin{bmatrix} 3 & 1 & -2 \\ -1 & 0 & -3 \\ -2 & 6 & 6 \end{bmatrix} = 3(18)-1(-6-6)-2(-6)$$

$$= 78$$

$$I_2 = \frac{\Delta_2}{\Delta} = \frac{78}{39} \Rightarrow 2A.$$

To find I_3:

$$\Delta_3 = \begin{bmatrix} 3 & -1 & 1 \\ -1 & 6 & 0 \\ -2 & -3 & 6 \end{bmatrix} = 3(36)-(-1)(-6)+1(3+12)$$

$$= 117$$

$$I_3 = \frac{\Delta_3}{\Delta} = \frac{117}{39} \Rightarrow 3A.$$

5. Let us determine the voltage across 5Ω resistance and the current in the 12V source by using Nodal analysis.

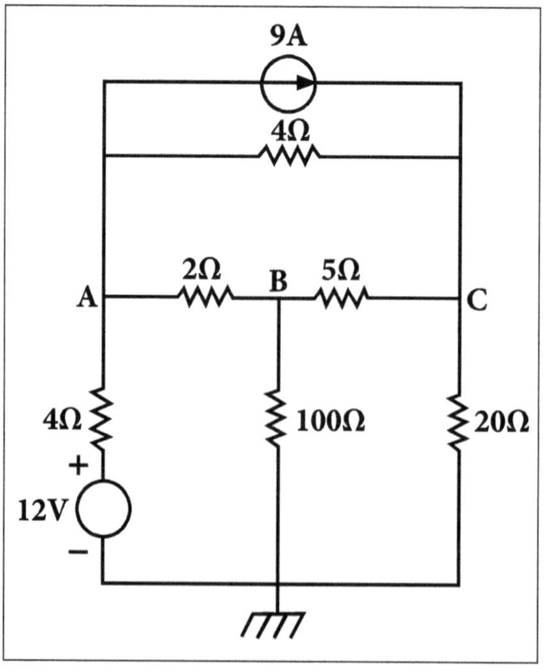

Solution:

Given:

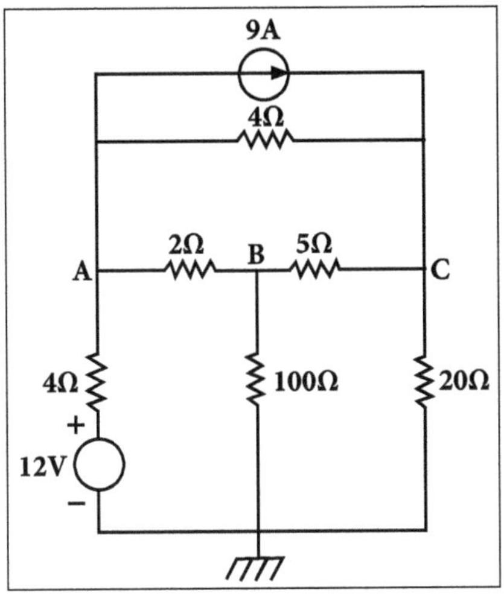

Using Nodal Method:

At Node A:

$$\frac{V_A - V_B}{2} + \frac{V_A - 12}{4} + \frac{V_A - V_C}{4} = -9$$

$$V_A\left(\frac{1}{2} + \frac{1}{4} + \frac{1}{4}\right) - \frac{V_B}{2} - \frac{V_C}{4} - \frac{12}{4} = -9$$

$$V_A - \frac{V_B}{2} - \frac{V_C}{4} = -6$$

$$4V_a - 2V_B - V_C = -24 \qquad \text{...(1)}$$

At Node B:

$$\frac{V_B - V_A}{2} + \frac{V_B}{100} + \frac{V_B - V_C}{5} = 0$$

$$-V_A + V_B\left(\frac{1}{2} + \frac{1}{100} + \frac{1}{5}\right) - \frac{V_C}{5} = 0$$

$$-\frac{V_A}{2} + V_B\left(\frac{H}{100}\right) - \frac{V_C}{5} = 0$$

$$-50V_A + 71V_B - 20V_C = 0 \qquad \text{...(2)}$$

At Node C:

$$\frac{V_C - V_B}{5} + \frac{V_C}{20} + \frac{V_C - V_A}{4} = 9$$

$$-\frac{V_A}{4} - \frac{V_B}{5} + V_C\left(\frac{1}{5} + \frac{1}{20} + \frac{1}{4}\right) = 9$$

$$-\frac{V_A}{4} - \frac{V_B}{5} + V_C\left(\frac{10}{20}\right) = 9$$

$$-5V_A - 4V_B + 10V_C = 180 \qquad \text{...(3)}$$

To find V_A, V_B and V_C:

$$\Delta = \begin{vmatrix} 4 & -2 & -1 \\ -50 & 71 & -20 \\ -5 & -4 & 10 \end{vmatrix} = 4(710+80) + 2(-500-100) - 1(200+355)$$

$$= 2520 - 1200 - 555$$

$$= 765$$

$$\Delta_B = \begin{vmatrix} 4 & -24 & -1 \\ \multicolumn{3}{c}{-\text{üüüüüüüüüüüü}} \\ -5 & 180 & 10 \end{vmatrix} = (\ +\ ) + (-\ -\ ) - (-\ +\ )$$

$$= 14400 - 14400 + 9000$$

$$= 9000$$

$$V_B = \frac{\Delta_B}{\Delta} = \frac{9000}{765} = 11.764 \text{ volts}$$

$$\Delta_C = \begin{vmatrix} 4 & -2 & -24 \\ -50 & 71 & 0 \\ -5 & -4 & 180 \end{vmatrix} = 4(12780+0) + 2(-9000+0) - 24(200+355)$$

$$= 51120 - 18000 - 13320$$

$$= 19800$$

$$\therefore \quad V_C = \frac{\Delta_C}{\Delta} = \frac{19800}{765} = 25.88\text{ V.}$$

∴ Current through 5Ω resistor is:

$$\frac{V_B - V_C}{5} = \frac{11.76 - 25.88}{5} = -14.12 \text{ Amps.}$$

1.3.1 Star-Delta Conversion

Star Connection

Star connection or star network is a network of three electrical components in a composition such that one point of all the components or resistors is connected at a common point. The star connection of three resistors or branches also looks like the English alphabet “Y”. Hence, the Star network or connection is also referred to as “Y” or “Wye” connection.

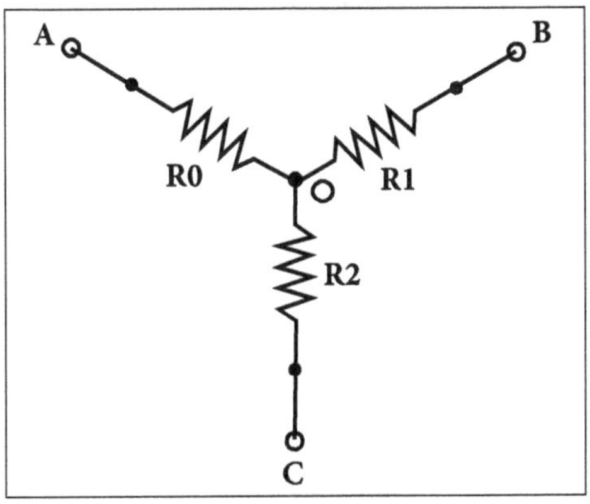

Star Network.

A star or Y connection can also be rearranged to make it look like the English letter “T” and many electrical circuits use the star connection in its “T” form.

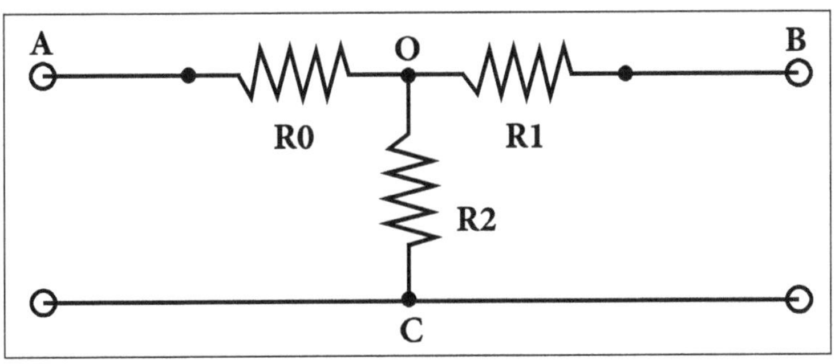

Star-T Network.

Delta Connection

Delta connection is an electrical network composed of three branches or circuit components, where each branch or element is connected head to tail of other branch or circuit element. The Delta connection is so called because it looks like the Greek alphabet Delta (Δ).

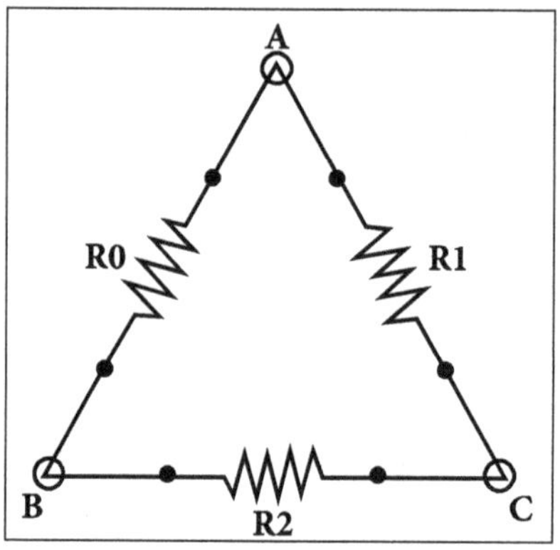

Delta Network.

The delta connection can also be rearranged to look like a Pi (π) symbol so the network is also sometimes referred to as Pi Network.

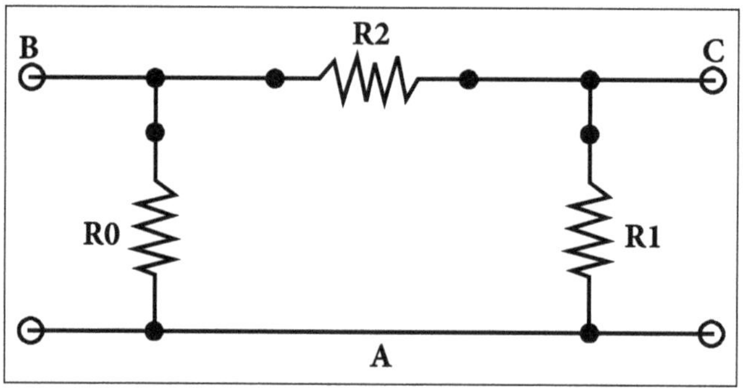

Delta (Pi(π) network).

It is often useful to transform delta network to star and star network to delta while solving an electrical network. The transformation can make the calculations simpler and easier.

Derivation for Star to Delta Conversion

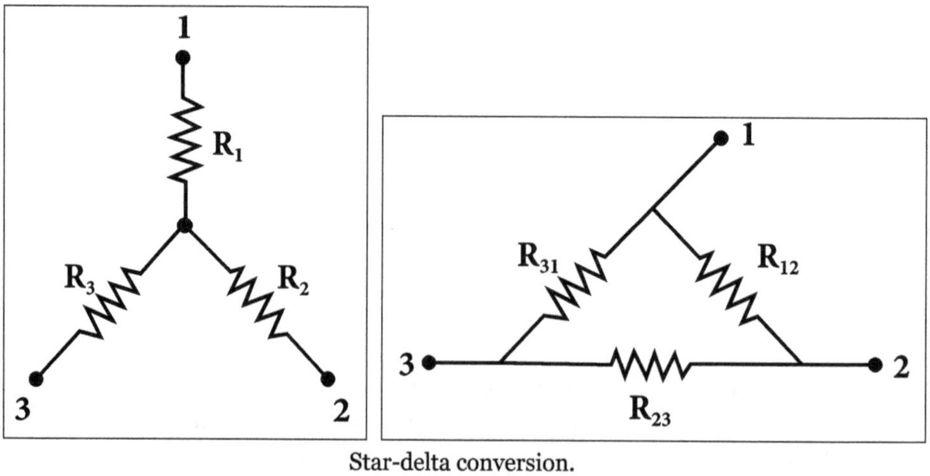

Star-delta conversion.

For the two circuits to be equivalent, the resistance between any two points with the third open should be equal for both the circuits.

∴ Between terminals (1) and (2) with (3) open:

$$R_1 + R_2 = R_{12} \,||\, (R_{23} + R_{31})$$

$$\therefore R_1 + R_2 = \frac{(R_{12})(R_{23} + R_{31})}{(R_{12} + R_{23} + R_{31})} \qquad ...(1)$$

Between terminals (2) and (3) with (1) open:

$$R_2 + R_3 = \frac{(R_{23})(R_{12} + R_{31})}{R_{12} + R_{23} + R_{31}}$$

$$R_2 + R_3 = R_{23} \,||\, (R_{12} + R_{31}) \qquad ...(2)$$

Between terminals (3) and (1) with (2) open:

$$R_3 + R_1 = R_{31} \,||\, (R_{12} + R_{23})$$

$$R_3 + R_1 = \frac{R_{31}(R_{12} + R_{23})}{R_{12} + R_{23} + R_{31}} \qquad ...(3)$$

By solving (1) - (2) + (3), we get:

$$2R_1 = \frac{\cancel{R_{12}R_{23}} + R_{12}R_{31} - \cancel{R_{12}R_{23}} - \cancel{R_{23}R_{31}} + R_{12}R_{31} + \cancel{R_{23}R_{31}}}{R_{12} + R_{23} + R_{31}}$$

$$\therefore \cancel{2}R_1 = \frac{\cancel{2}R_{12}R_{31}}{R_{12} + R_{23} + R_{31}}$$

$$\therefore R_1 = \frac{R_{12}\, R_{31}}{R_{12} + R_{23} + R_{31}} \qquad ...(4)$$

Similarly,

$$R_2 = \frac{R_{12}\, R_{23}}{R_{12} + R_{23} + R_{31}} \qquad ...(5)$$

And,

$$R_3 = \frac{R_{23}\, R_{31}}{R_{12} + R_{23} + R_{31}} \qquad ...(6)$$

Thus, equations (4), (5) and (6) will be used for delta to star conversion.

Now, dividing equation (6) by equation (5), we get:

$$\frac{R_3}{R_2} = \frac{R_{31}}{R_{12}}$$

$$\therefore R_{31} = R_{12}\left[\frac{R_3}{R_2}\right] \qquad ...(7)$$

Dividing equation (6) by equation (4), we get:

$$\frac{R_3}{R_1} = \frac{R_{23}}{R_{12}}$$

$$\therefore \quad R_{23} = R_{12}\left[\frac{R_3}{R_1}\right] \qquad ...(8)$$

Substituting equation (7) and (8) in equation (4), we get:

$$R_1 = \frac{R_{12}.R_{12}(R_3/R_2)}{R_{12} + R_{12}(R_3/R_1) + R_{12}(R_3/R_2)} = \frac{R_{12}^{\not 2}(R_3/R_2)}{\not R_{12}(1 + R_3/R_1 + R_3/R_2)}$$

$$\therefore R_{12} = R_1\left[\frac{R_2}{R_3}\right]\left[1 + \frac{R_3}{R_1} + \frac{R_3}{R_2}\right]$$

$$\therefore R_{12} = \left[\frac{R_1 R_2}{R_3}\right]\left[1 + \frac{R_3}{R_1} + \frac{R_3}{R_2}\right]$$

$$\therefore R_{12} = R_1 + R_2 + \frac{R_1 R_2}{R_3} \qquad ...(9)$$

Similarly,

$$R_{23} = R_2 + R_3 + \frac{R_2 R_3}{R_1} \qquad ...(10)$$

And,

$$R_{31} = R_3 + R_1 + \frac{R_3 R_1}{R_2} \qquad ...(11)$$

Thus, equations (9), (10) and (11) will be used for star to delta conversion.

Delta (Δ) - Wye Conversion (Y)

These configurations may often handle by the use of a Δ-Y or Y-Δ transformation. One of the most basic three-terminal network equivalent is that of three resistors connected in "Delta (Δ)" and in "Wye (Y) ".

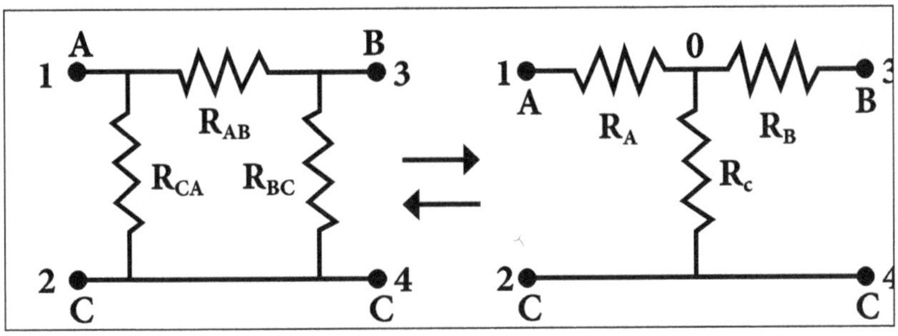

(a) and (b) Four-terminal networks.

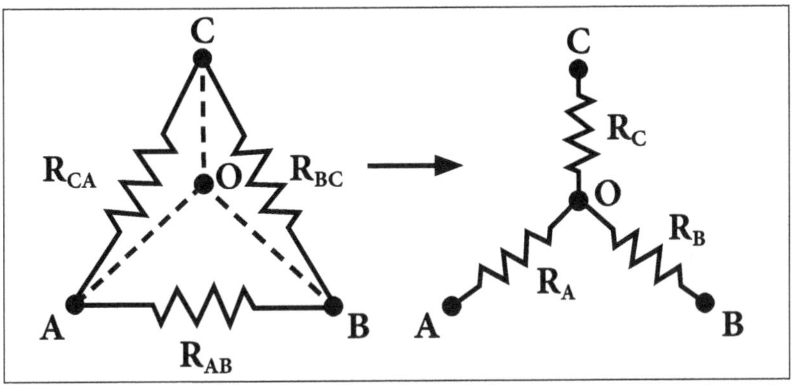

(c) Three terminal networks.

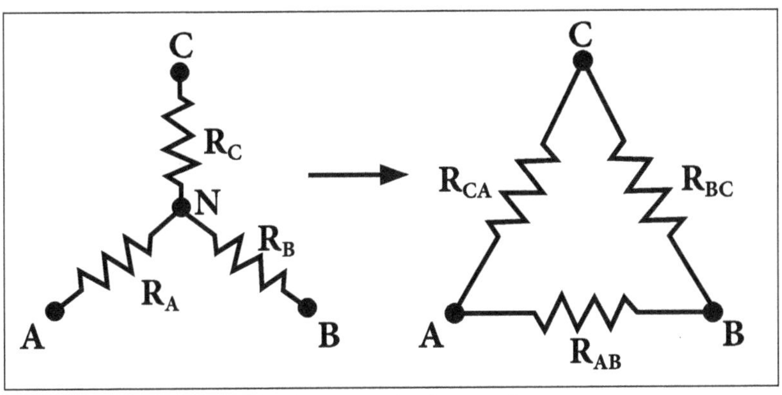

(d) Three terminal networks.

These two circuits identified in figure (c) and figure (d) are sometimes part of a larger circuit and obtained their names from their configurations. These three terminal networks can be redrawn as four-terminal networks as shown in figure (a) and figure (b).

We can obtain useful expression for direct transformation or conversion from Δ to Y or Y to Δ by considering that the two networks have the same resistance when looked at the similar pairs of terminals.

Conversion from Delta (Δ) to Star or Wye (Y)

Let us consider the network shown in figure (c) and assumed the resistances (R_{AB}, R_{BC} and R_{CA}) in Δ network are known. Our problem is to find the values of R_A, R_B and R_C in Wye (Y) network figure (a) that will produce the same resistance when measured between similar pairs of terminals.

We can write the equivalence resistance between any two terminals in the following form:

Between A & C terminals:

$$R_A + R_C = \frac{R_{CA}(R_{AB} + R_{BC})}{R_{AB} + R_{BC} + R_{CA}} \qquad \text{...(1)}$$

Between C & B terminals:

$$R_C + R_B = \frac{R_{BA}(R_{AB} + R_{CA})}{R_{AB} + R_{BC} + R_{CA}} \qquad \text{...(2)}$$

Between B & A terminals:

$$R_B + R_A = \frac{R_{AB}(R_{CA} + R_{BC})}{R_{AB} + R_{BC} + R_{CA}} \qquad \text{...(3)}$$

By combining above three equations, we can write an expression as given below:

$$R_A + R_B + R_C = \frac{R_{AB}R_{BC} + R_{BC}R_{CA} + R_{CA}R_{AB}}{R_{AB} + R_{BC} + R_{CA}} \qquad \text{...(4)}$$

Subtracting equations (2), (1) and (3) from equation (4), we can write the expression for unknown resistances of Wye (Y) network as:

$$R_A = \frac{R_{AB}R_{AC}}{R_{AB} + R_{BC} + R_{CA}} \qquad \text{...(5)}$$

$$R_B = \frac{R_{AB}R_{BC}}{R_{AB} + R_{BC} + R_{CA}} \qquad \text{...(6)}$$

$$R_C = \frac{R_{BC}R_{CA}}{R_{AB} + R_{BC} + R_{CA}} \qquad \text{...(7)}$$

Problems

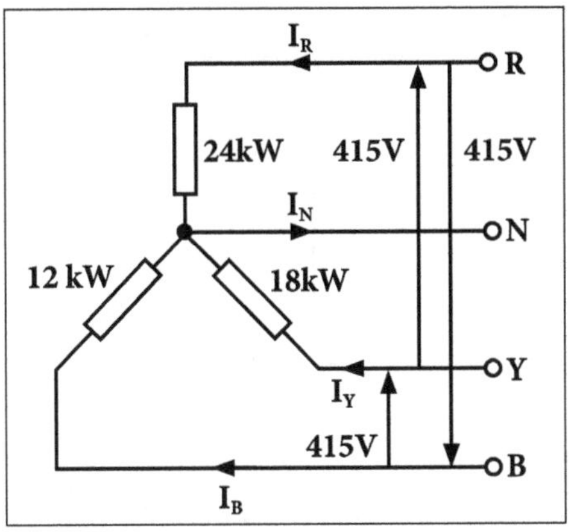

1. A 415V, 3-phase, 4 wire, star-connected system supplies three resistive loads as shown in figure below. Let us determine: (a) the current in each line and (b) the current in the neutral conductor.

Solution:

Given:

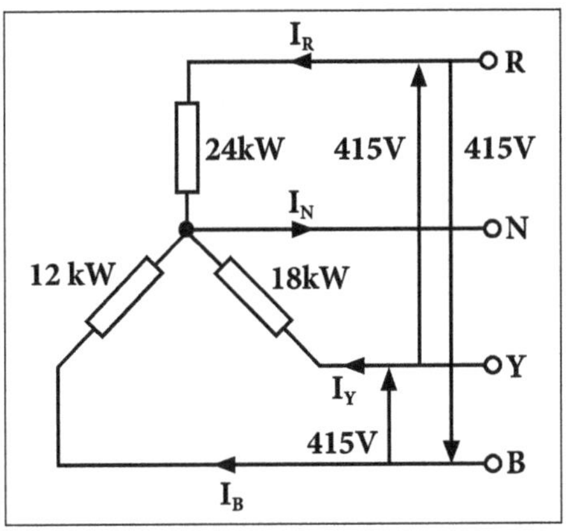

A 415V, 3-phase, 4 wires, star-connected system supplies three resistive loads.

(a) For a star-connected system, $V_L = \sqrt{3}V_p$

Hence, $V_P = \frac{V_L}{\sqrt{3}} = \frac{415}{\sqrt{3}} = 240\text{V}$

Since, Current is I = Power P/Voltage V for a resistive load.

Then,

$$I_R = P_R / V_R = 24000/240 = 100\text{A}$$

$$I_Y = P_Y / V_Y = 18\,000/240 = 75\text{A}$$

$$I_B = P_B / V_B = 12\,000/240 = 50A$$

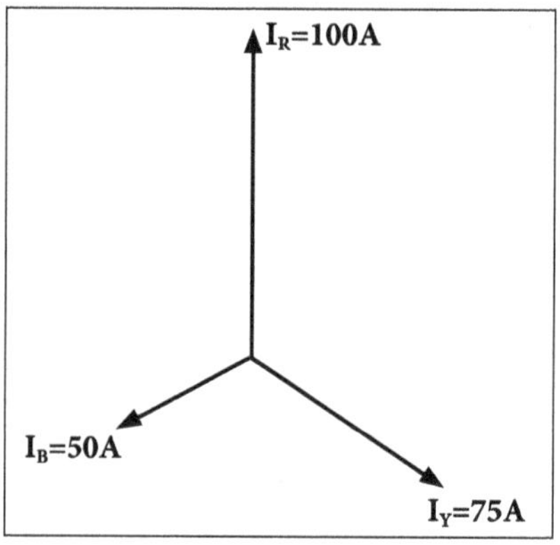

(b) The three line currents are shown in the phasor diagram below. Since each load is resistive, the currents are in phase with the phase voltages and are mutually displaced by 120°.

The current in the neutral conductor is given by:

$$I_N = I_R + I_Y + I_B \text{ phasorially.}$$

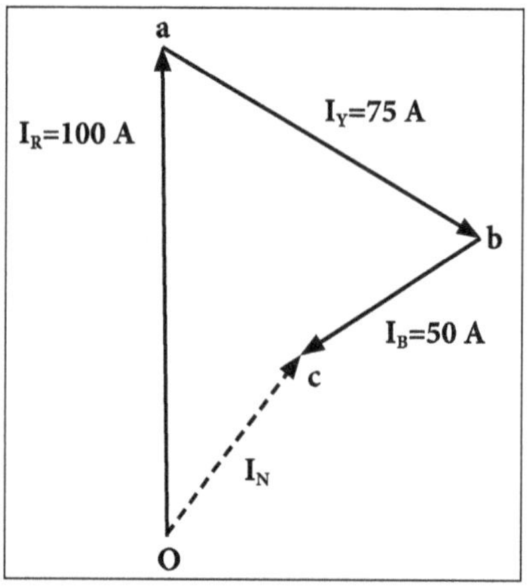

The above figure shows the three line currents added phasorially, oa represents I_R in magnitude and direction. From the nose of oa, ab is drawn representing I_Y in magnitude and direction. From the nose of ab, bc is drawn representing I_B in its magnitude and direction oc represents the resultant, I_N.

By measurement, $I_N = 43A$

Alternatively, by calculation, considering I_R at 90°, I_B at 210° and I_Y at 330°:

Total horizontal component $= 100\cos 90° + 75\cos 330° + 50\cos 210° = 21.65$

Total vertical component $= 100\sin 90° + 75\sin 330° + 50\sin 210° = 37.50$

Hence, magnitude of $I_N = \sqrt{\left(21.65^2 + 37.50^2\right)} = 43.3A$

2. Let us convert the given delta figure into equivalent star.

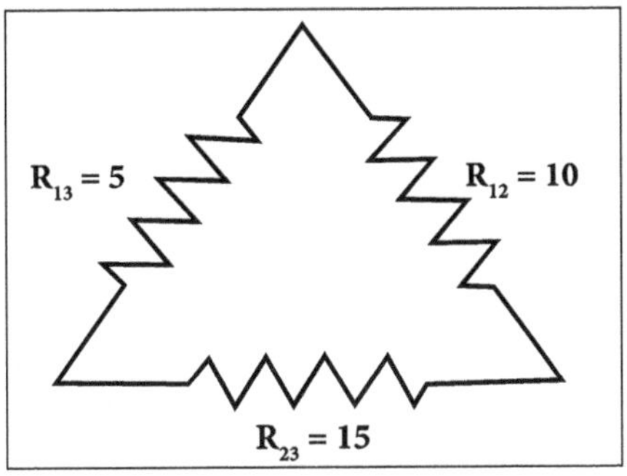

Solution:

Given:

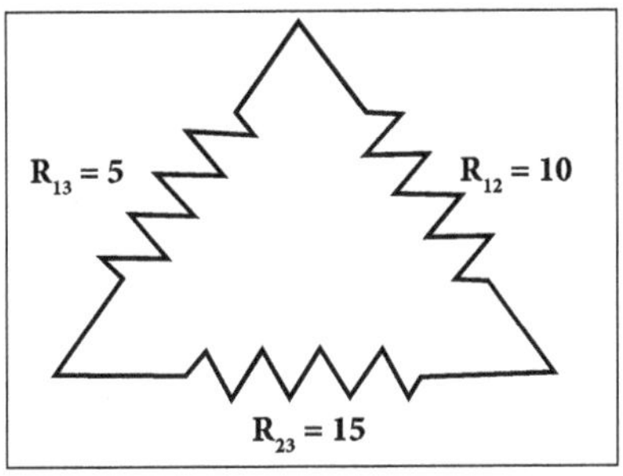

$$R_1 = R_{12} \times R_{13} / R_{12} + R_{13} + R_{23}$$

$$R_1 = 10 \times 5/30 = 1.67\Omega$$

$$R_2 = 10 \times 15/30 = 5\Omega$$

$$R_3 = 5 \times 15/30 = 2.5\Omega$$

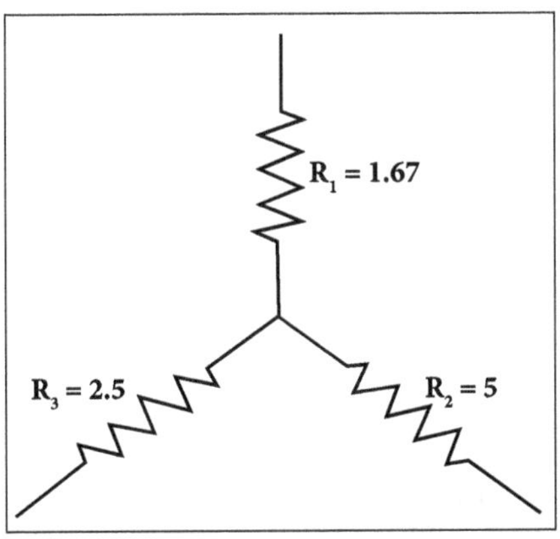

3. Consider that the below figure transform delta to star or Wye circuit where the values R_1 = 20 ohms, R_2 = 30 ohms and R_3 = 50 ohms. Let us determine the value of R_a, R_b and R_c.

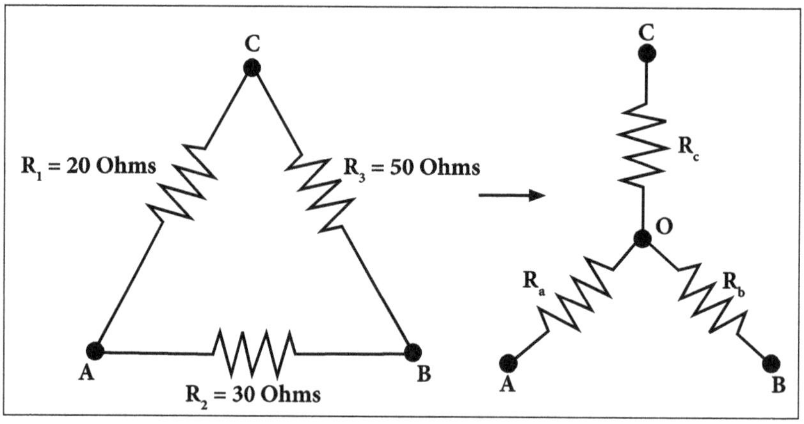

Solution:

Given:

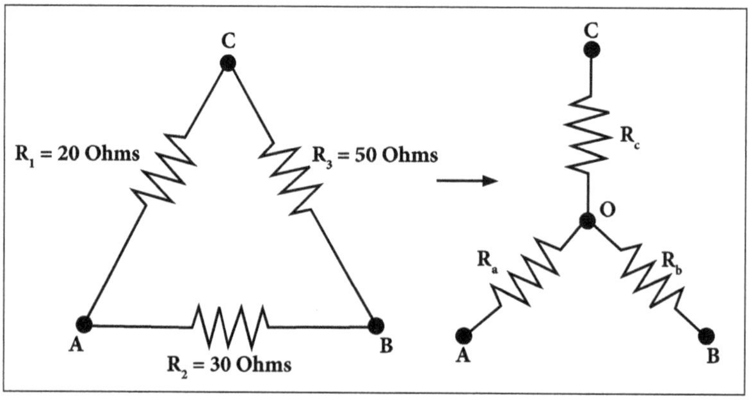

For delta to star conversions, equivalent resistance equations are:

$$R_a = (R_1 \; R_2)/(R_1 + R_2 + R_3)$$

$$R_b = (R_2 \; R_3)/(R_1 + R_2 + R_3)$$

$$R_c = (R_1 \; R_3)/(R_1 + R_2 + R_3)$$

Therefore, the total resistance, $R_t = (R_1 + R_2 + R_3)$

$$= 20 + 30 + 50$$

$$= 100 \text{ ohms}$$

$$R_a = (R_1 \; R_2)/(R_1 + R_2 + R_3)$$

$$=(20\times 30)/100$$

$$=6 \text{ ohms}$$

Similarly, $R_b=(R_2\ R_3)/(R_1+R_2+R_3)$

$$=(30\times 50)/100$$

$$=15 \text{ ohms}$$

And, $R_c=(R_1\ R_3)/(R_1+R_2+R_3)$

$$=(50\times 20)/100$$

$$=10 \text{ ohms}$$

4. Consider the below figure to transform star or Wye to the delta circuit where the resistance values in star network are given as R_1 = 10 ohms, R_2 = 5 ohms and R_3 = 20 ohms. Let us determine the value of R_{12}, R_{23} and R_{31}.

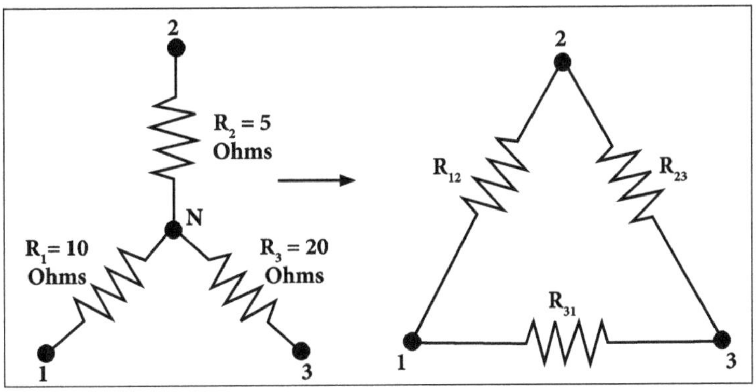

Solution:

Given:

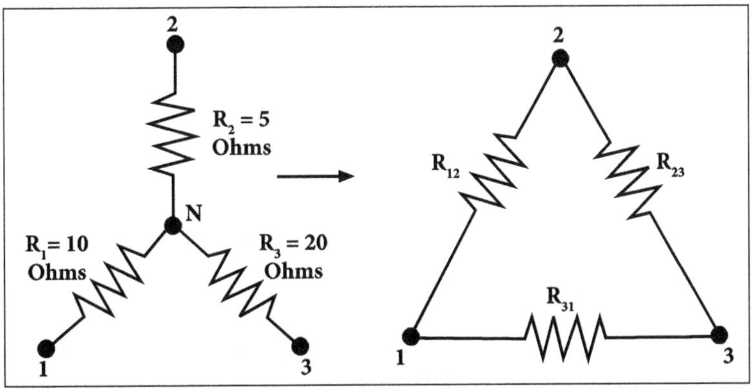

For star to delta conversion, the equivalent resistance equations are as follows:

$$R_{12} = R_1 + R_2 + ((R_1 R_2)/R_3)$$

$$R_{23} = R_2 + R_3 + ((R_2 R_3)/R_1)$$

$$R_{31} = R_1 + R_3 + ((R_1 R_3)/R_2)$$

By simplifying the above equations, we get the common numerator term as:

$$R_1 R_2 + R_2 R_3 + R_1 R_3 = 10 \times 5 + 10 \times 20 + 20 \times 5$$

$$= 350 \text{ ohms}$$

Then $R_{12} = 350/R_3$

$$= 350/20$$

$$= 17.5 \text{ ohms}$$

$$R_{23} = 350/R_1$$

$$= 350/10$$

$$= 35 \text{ ohms}$$

$$R_{31} = 350/R_2$$

$$= 350/5$$

$$= 70 \text{ ohms}$$

1.3.2 Maximum Power Transfer Theorem with Proof

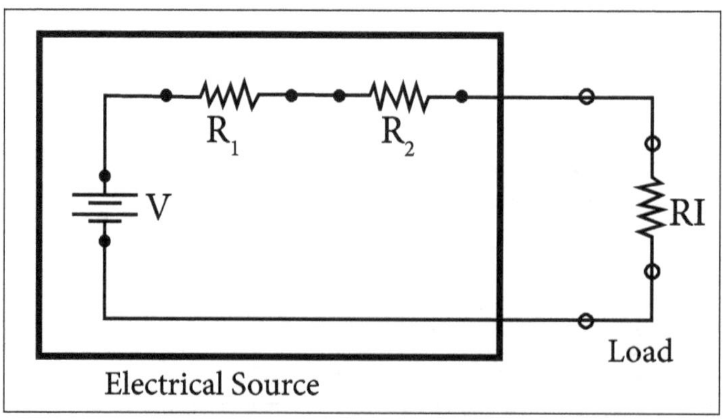

Maximum power transfer theorem.

Maximum power theorem is a theorem or technique used in the Electrical Network Analysis and Electrical circuit designing. It was invented by a German engineer Moritz von Jacobi in the year 1840. He invented the theorem in the process of finding a way to maximize the output of the battery to a motored boat which he designed to travel in the river Neva. Hence, the theorem is also sometimes referred as Jacobi's Law.

Maximum power transfer theorem deals with the power transferred to the load on a circuit with a network of various sources or components on it. The maximum power transfer theorem defines the condition under which the maximum power is transferred to the load in a circuit.

Under the condition of Maximum power transfer, we only deal with the power transferred to the load and does not consider the power dissipated in internal circuits or resistance of the source. Thus, we are not talking about the maximum efficiency of power transfer, but instead maximum possible power transfer from a source to a load.

According to maximum power transfer theorem, maximum power is yielded to the load R_L when R_L is equal to the internal resistance of the circuit or R_1+R_2. The Maximum power transfer theorem is true in case of any kind of circuit if it may be linear, non-linear, active, DC or AC.

In case of DC circuits, load resistance is matched with the internal resistance of the source by making both resistance equal and in case of AC, the load impedance is matched with the internal impedance of the circuit or source by making the load impedance as the complex conjugate of the source impedance.

For example, load impedance will be R_1 - jX if the internal impedance of the source is R_1+ jX.

Proof of Maximum Power Transfer Theorem

Let,

V = EMF supplied to the load.

R_L = Load resistance.

R_i = Internal resistance of the source.

I = Current flowing through the load, internal resistance and the source of the circuit.

P_L = Power transferred to the load.

P_i = Power dissipated at the internal resistance.

Then,

Power transferred to the load $= P_L = I^2 R_L$

$$P_L = \left(\frac{V}{R_i + R_L}\right)^2 \times R_L = \frac{V^2}{\frac{R_i^2}{R_L} + 2R_i + R_L}$$

Now using the theorems of Differential calculus, if we keep the value of R_L as variable and want to calculate the maximum value of P_L, then we need to differentiate P_L with respect to R_L and equate it with zero. Thus, under Maximum power transfer to load condition:

$$\frac{d}{dR_L} P_L = \frac{d}{dR_l} \frac{V^2}{\frac{R_i^2}{R_L} + 2R_i + R_L} = 0$$

Or,

$$-\frac{R_i^2}{R_L^2} + 1 = 0$$

Or,

$$R_i = R_L$$

And in AC networks using same mathematical technique, we can prove.

If,

$Z_i = R_i + X_i =$ Internal impedance of reactive circuits.

$Z_L = R_L + X_L =$ Load impedance.

Then, under the condition of maximum power transfer to load:

$$Z_i = Z_l,\ R_i = R_L \text{and } X_i = -X_L$$

Power Transfer Efficiency

Power transfer efficiency is the efficiency of any source or circuit in transferring its power to the load or it is the ratio of power transferred to the load over total power transferred by the source. It is denoted by the Greek letter η.

Mathematically:

$$\eta = P_L / P_T$$

Where,

P_L is the power transferred to the load.

P_T is the total power transferred by the source.

We can expand this expression as,

$$P_L = I^2R_L$$

$$P_T = P_L + P_I = I^2R_L + I^2R_I$$

Where,

I is the total current flowing through the circuit or Thevenin's equivalent of the circuit.

P_I is the power dissipated in internal circuits of the source.

$$\eta = \frac{P_L}{P_T} = \frac{I^2R_L}{I^2R_L + I^2R_I} = \frac{R_L}{R_L + R_I} = \frac{1}{1 + \frac{R_I}{R_L}}$$

Thus,

As we know that under the condition of maximum power transfer $R_L = R_I$, we can derive that the efficiency under the condition of maximum power transfer is only 0.5.

The overall efficiency decreases if the value of R_L is kept very low and it increases up to one when the value of R_L is increased to infinity as shown on the graph below and the power transferred to load becomes minimum when R_L is kept very low and very high.

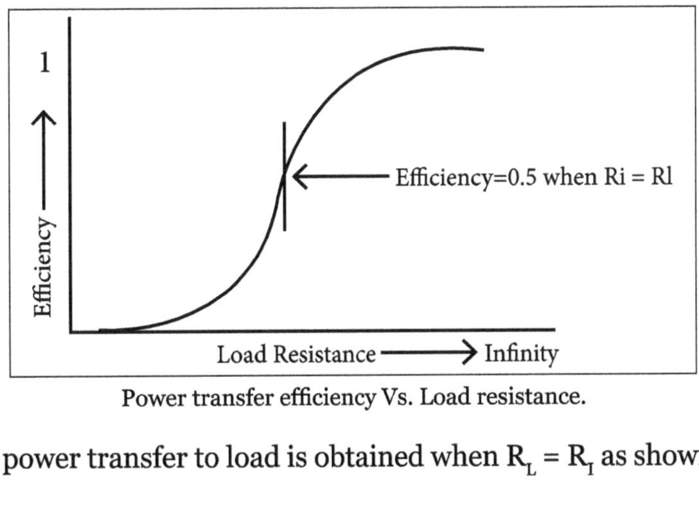

Power transfer efficiency Vs. Load resistance.

The maximum power transfer to load is obtained when $R_L = R_I$ as shown on the following graphs.

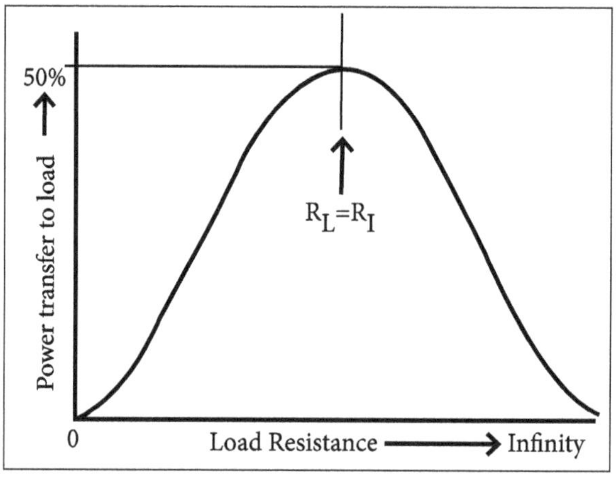

Power transfer to load Vs. Load Resistance.

Applications of Maximum Power Transfer Theorem

The Maximum Power Transfer Theorem has a wide range of usage on real life situation. This theorem is used to maximize the power output to a load from any circuit. So, they can be used to design circuits where the maximum output performance is desired for example to match the amplifier with the loudspeaker to yield the maximum power to the speaker and therefore, it produces maximum sound.

In some situations, Transformer Coupling are also used to yield maximum power to the load when the matching of Load and Source impedance is not possible. The application of Maximum Power Theorem is done only under the conditions when the maximum performance is desired over the overall efficiency of the circuit because the efficiency of the circuit under maximum power transfer condition is only 0.5.

Problems

1. If an 8Ω loudspeaker is to be connected to an amplifier with an output impedance of 1000Ω, let us calculate the turns ratio of the matching transformer required to provide maximum power transfer of the audio signal. Assume that the amplifier source impedance is Z_1, the load impedance is Z_2 with the turns ratio given as N.

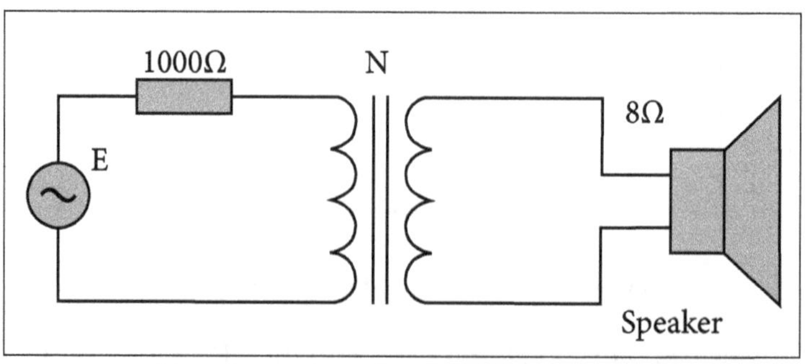

Transformer impedance matching circuit.

Solution:

Given:

8Ω loudspeaker

Output impedance = 1000Ω

Amplifier source impedance = Z_1

Load impedance = Z_2

Transformer turns ratio:

$$Z_1 = N^2 Z_2 \qquad \therefore \ N = \sqrt{\frac{Z_1}{Z_2}}$$

Therfore,

$$N = \sqrt{\frac{Z_1}{Z_2}} = \sqrt{\frac{1000}{8}} = 11.2:1$$

Generally, small transformers used in low power audio amplifiers are usually regarded as ideal. So, any losses can be ignored.

2. Let us determine the value of R_L for the given network below that the power is maximum. And also determine the Max Power through load-resistance R_L by using maximum power transfer theorem.

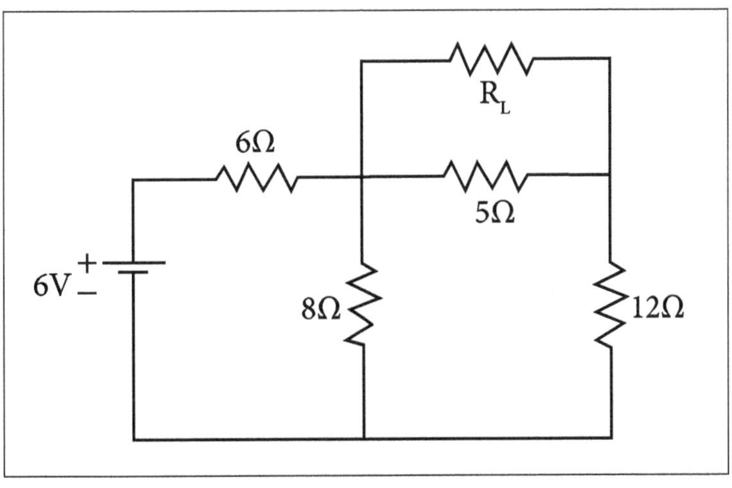

Solution:

Given:

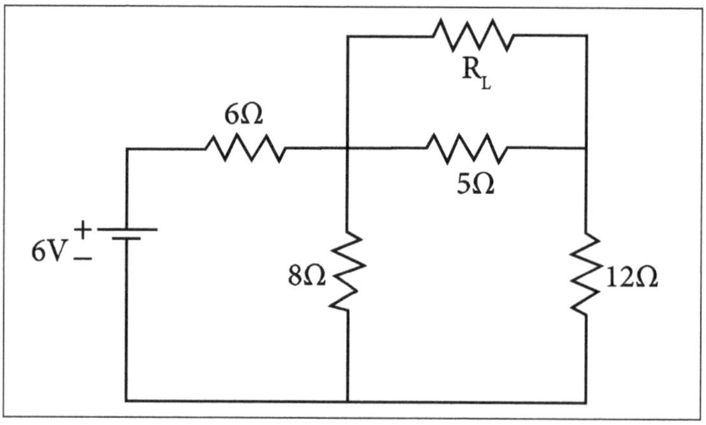

For the above network, we are going to determine the value of unknown resistance called "R_L".

When power is maximum through load, resistance is equal to the equivalent resistance between the two ends of load-resistance after removing.

So, for finding load-resistance R_L, we have to determine the equivalent resistance for this circuit.

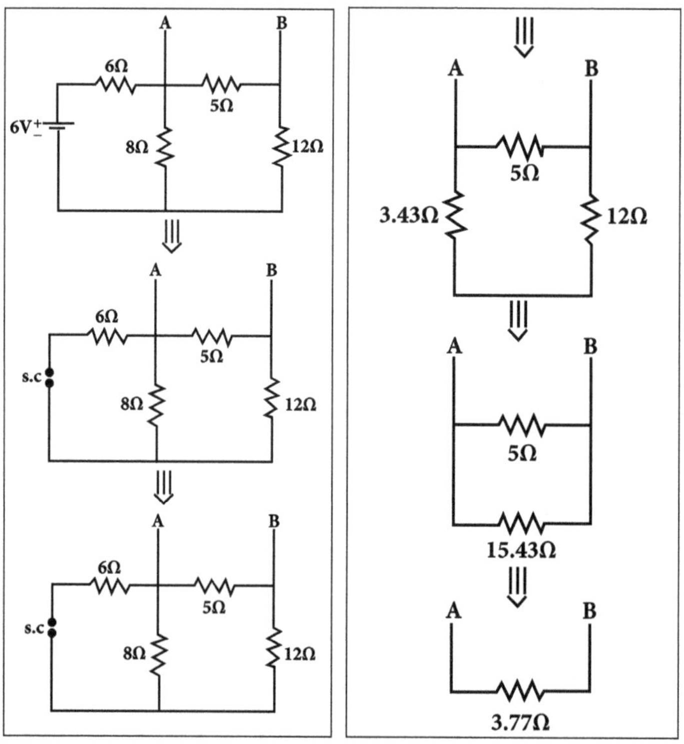

Therefore,

$$R_{AB} / R_L = 3.77\,\Omega$$

Now, for finding Maximum Power through load-resistance, we have to find out the value of $V_{o.c}$.

Here, $V_{o.c}$ is known as voltage between open circuits.

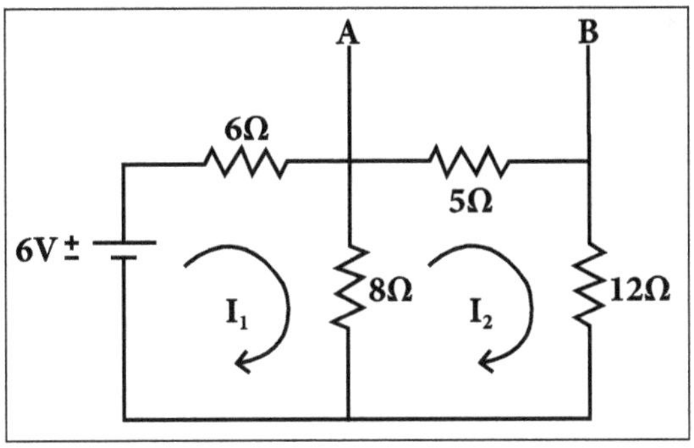

Applying KVL in 1st loop, we get:

$$6 - 6I_1 - 8I_1 + 8I_2 = 0$$

$$-14I_1 + 8I_2 = -6 \qquad ...(1)$$

Again, Applying KVL in 2nd loop, we get:

$$-8I_2 - 5I_2 - 12I_2 + 8I_1 = 0$$

$$8I_1 - 25I_2 = 0 \qquad ...(2)$$

On solving equations (1) and (2), we get:

$$I_1 = 0.524\text{ A}$$

$$I_2 = 0.167\text{ A}$$

Now, from the circuit $V_{o.c}$ is:

$$V_A - 5I_2 - V_B = 0$$

$$V_{o.c} / V_{AB} = 5I_2$$

$$= 5 \times 0.167$$

$$= 0.835\text{V}$$

So, the maximum power through R_L is given by:

$$P_{max} = \frac{V_{o.c}^2}{4R_L}$$

$$P_{max} = \frac{0.835^2}{4 \times 3.77}$$

$$P_{max} = 0.046 \text{ watt}.$$

2

Electromagnetism

2.1 Biot-Savart Law

According to Biot-Savart's law, the magnetic field at a point P due to an incremental element of length i.e., dl carrying a current is directly proportional to length of the element, current carried by the element, sine of the angle θ between the line joining the point P with the length i.e., dl and inversely proportional to the square of the distance of this point from the element.

Hence, field strength dH is given by:

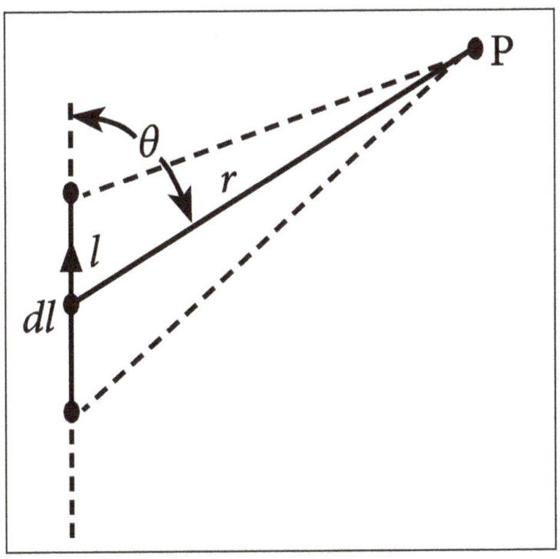

Biot-Savart's law.

$$dH \propto \frac{Idl}{r^2}\sin\theta$$

$$dH = \frac{1}{4\pi}\frac{Idl}{r^2}\sin\theta$$

$$dH = \frac{Idl\sin\theta}{4\pi r^2} A/m$$

Where,

dH = Magnetic field at a point, P.

Idl = Differential current element (A-m).

ar = Unit vector along the line joining the point P and the Idl.

r = Distance of P from the current element (m).

$1/4\pi$ = Constant of proportionality.

Biot-savart's law is very helpful in calculating the field strength H due to a particular configuration of conductor carrying a current.

The direction of dH is normal to the plane containing Idl and the unit vector along the line from the element to the point. This normal is in the direction of progress of a right-handed screw turned from dl through a small angle to the line from the element to the point.

The direction of the magnetic field is easily known using the right hand thumb rule. The magnetic field is in closed loops around the current element.

If a current element is held in the right hand with the thumb pointing upwards indicating the direction of current, then the direction of the remaining fingers indicate the direction of the magnetic field, that is, if the current is upwards, the direction of magnetic field is anti-clockwise and if the current is downwards, the direction of magnetic field is clockwise.

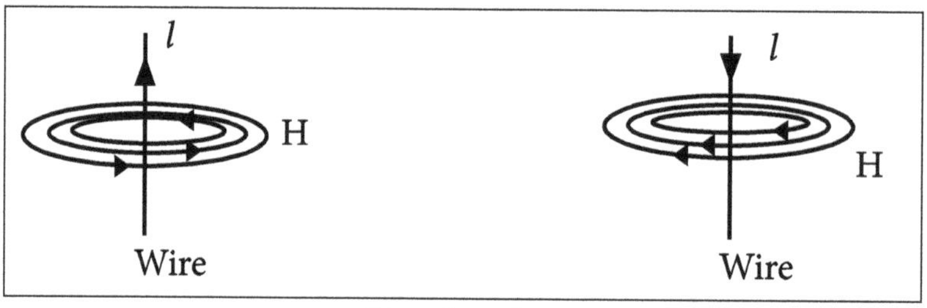

Direction of H.

2.1.1 Ampere's Circuital Law

Ampere's Law

The line integral of the magnetic field around a closed path is equal to the surface integral of current density of the same loop.

$$\oint_C \vec{H}.d\vec{l} = \int_S \vec{J}.d\vec{a} \quad ...(1)$$

Where,

$\vec{H}$ - Magnetic field intensity vector.

$\vec{J}$ - Current density.

Magnetic flux crossing surface S:

$$\oint_S \vec{B}.d\vec{a} = 0 \qquad ...(2)$$

Where,

$\vec{B}$ - Magnetic flux density vector.

$$\vec{B} = \mu\vec{H} \qquad ...(3)$$

$$\mu = \mu_r \mu_o$$

μ - Magnetic permeability, Wb/A-t-m = H/m.

μ_o - The permeability of free space(4π.10−7).

μ_r - Relative permeability.

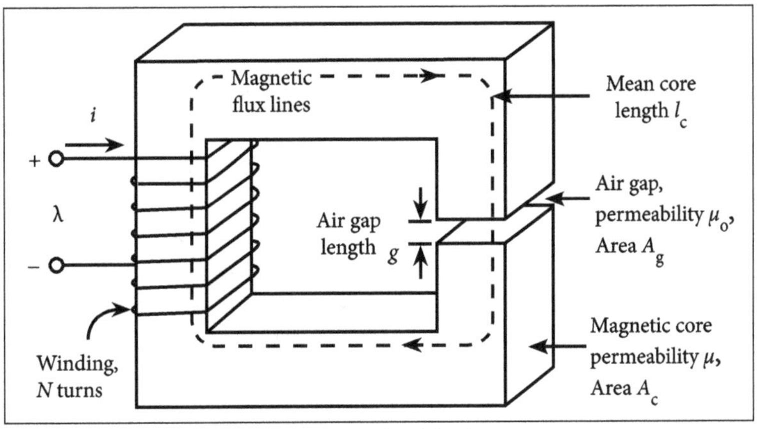

Magnetic circuit with air gap.

Flux is the same in the magnetic core and the air gap and is obtained as:

$B_c = \dfrac{\phi}{A_c}$ Flux density in the magnetic core.

$B_g = \dfrac{\phi}{A_g}$ Flux density in the air-gap.

mmf is given by:

$$\oint_C \vec{H}.d\vec{l} = H_c l_c + H_g g = Ni = F \;\Rightarrow\; F = \frac{B_c}{\mu} l_c + \frac{B_g}{\mu_o} g = \phi\left(\frac{l_c}{\mu A_c} + \frac{g}{\mu_o A_g}\right)$$

$$\Rightarrow F = \phi\left(\Re_c + \Re_g\right) \Rightarrow \phi = \frac{F}{\Re_c + \Re_g}$$

Where,

$\mathfrak{R}_c$ - Reluctance of the core.

$\mathfrak{R}_g$ - Reluctance of the air gap.

Point form of Ampere's Circuital Law

Ampere's circuital law states that the line integral of H about any closed path is exactly equal to the direct current enclosed by that path.

$$\oint H.dl = I.$$

Let us consider a closed path 1, 2, 3, 4 in the Cartesian co-ordinate system.

H at the centre is $H = H_o = H_{xo}\, a_x + H_{yo}\, a_y + H_{zo}\, a_z$

We have four paths viz 1-2, 2-3, 3-4 and 4-1.

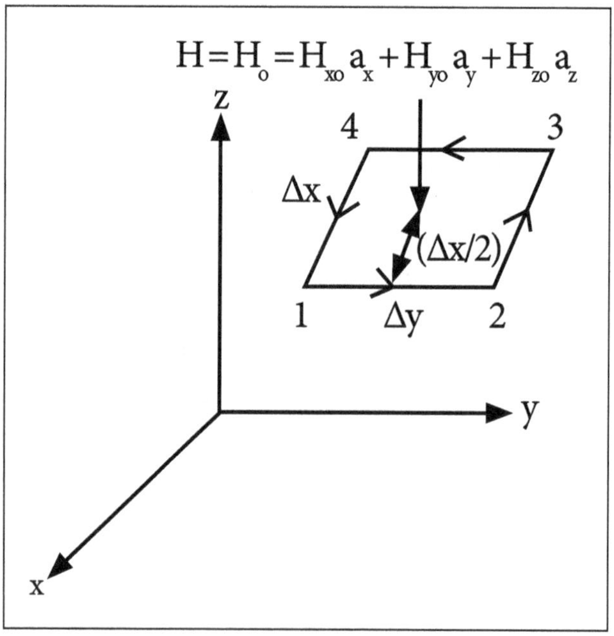

Point form of Ampere's Circuital Law.

$$(H.\Delta L)_{1-2} = H_{y1-2}\Delta y$$

$$= \left[H_{yo} + \frac{\partial H_y}{\partial x}\left(\frac{\Delta x}{2}\right)\right]\Delta y \qquad ...(1)$$

$$(H.\Delta L)_{2-3} = H_{x2-3}(-\Delta x) = \left[H_{xo} + \frac{\partial H_x}{\partial y}\left(\frac{\Delta y}{2}\right)\right](-\Delta x) \qquad ...(2)$$

$$(H.\Delta L)_{3-4} = H_{x3-4}(-\Delta y) = \left[H_{yo} + \frac{\partial H_y}{\partial x}\left(\frac{\Delta x}{2}\right)\right](-\Delta y) \quad ...(3)$$

$$(H.\Delta L)_{4-1} = H_{x4-1}(\Delta x) = \left[H_{xo} + \frac{\partial H_x}{\partial y}\left(\frac{-\Delta y}{2}\right)\right](\Delta x) \quad ...(4)$$

Adding (1), (2), (3) and (4):

$$\oint H.dL = \left[\frac{\partial H_x}{\partial x} - \frac{\partial H_y}{\partial y}\right]\Delta x.\Delta y$$

By Ampere's law, í $H.dL = \Delta I = \Delta x \Delta y J_z$

$$\oint H.dL = \left[\frac{\partial H_y}{\partial x} - \frac{\partial H_y}{\partial y}\right]\Delta x.\Delta y = \Delta x \Delta y J_z$$

As the closed path shrinks to zero:

$$\underset{\Delta y \Delta y \to o}{Lt} \frac{\oint H.dL}{\Delta x \Delta y} = \frac{\partial H_y}{\partial x} - \frac{\partial H_x}{\partial y} = J_z \quad ...(5)$$

Similarly,

$$\underset{\Delta y \Delta z \to o}{Lt} \frac{\oint H.dL}{\Delta y \Delta z} = \frac{\partial H_z}{\partial y} - \frac{\partial H_y}{\partial z} = J_x \quad ...(6)$$

$$\underset{\Delta z \Delta x \to o}{Lt} \frac{\oint H.dL}{\Delta z \Delta x} = \frac{\partial H_x}{\partial z} - \frac{\partial H_z}{\partial x} = J_y \quad ...(7)$$

Combining (5), (6) and (7):

$$\text{curl}\, H = \left(\frac{\partial H_z}{\partial y} - \frac{\partial H_y}{\partial z}\right)\hat{a}_x + \left(\frac{\partial H_x}{\partial z} - \frac{\partial H_z}{\partial x}\right)\hat{a}_y + \left(\frac{\partial H_y}{\partial x} - \frac{\partial H_x}{\partial y}\right)\hat{a}_z = J$$

$$\nabla \times H = \begin{vmatrix} a_x & a_y & a_z \\ \frac{\partial}{\partial x} & \frac{\partial}{\partial y} & \frac{\partial}{\partial z} \\ H_x & H_y & H_z \end{vmatrix} = J$$

This is the point form of Ampere's law.

2.1.2 Field Calculation: Biot-Savart and Ampere's Circuital Law

Magnetic Field Intensity (H)

The magnetic fields are generated by currents and calculated from Ampere's Law or the Biot-Savart Law are characterized by the magnetic field (B) measured in Tesla. But when the generated fields pass through magnetic materials which themselves contribute internal magnetic fields, the field comes from the external currents. It has been common practice to define another magnetic field quantity, usually known as the "magnetic field strength" designated by H. It can be defined by the relationship:

$$H = B_o / \mu_o$$

$$H = B / \mu_o - M$$

And has the value of unambiguously designating the driving magnetic influence from external currents in a material, independent of the material's magnetic response. The relationship for B can be written in the equivalent form:

$$B = \mu_o (H + M)$$

H and M has the same units, amperes per meter. To further distinguish B from H, B is sometimes called as the magnetic flux density or magnetic induction. The quantity M in these relationships is called the magnetization of the material.

Another commonly used form for the relationship between B and H is given by:

$$B = \mu_m H$$

Where,

$$\mu = \mu_m = K_m \mu_o$$

Where, μ_o is the magnetic permeability of space and Km is the relative permeability of the material. If the material will not respond to the external magnetic field by producing any magnetization, then K_m = 1. Another commonly used magnetic quantity is the magnetic susceptibility which specifies how much the relative permeability differs from one.

Magnetic susceptibility $\chi_m = K_m - 1$.

For paramagnetic and diamagnetic materials, the relative permeability is very close to 1 and the magnetic susceptibility is quite close to zero. For ferromagnetic materials, these quantities may be quite large.

The unit for the magnetic field strength H can be derived from its relationship to the magnetic field B, B = μH. Since the unit of magnetic permeability μ is N/A^2, then the unit for the magnetic field strength is given by:

$$T/\left(N/A^2\right)=(N/Am)/\left(N/A^2\right)=A/m$$

An older unit for magnetic field strength is the oersted: 1 A/m = 0.01257 oersted.

Expression for the Vector Magnetic Potential due to the Current Element and the Magnetic field Intensity at the Point (ρ, φ, z):

A differential current element Idz a_z is located at the origin in the free space. Let us obtain the expression for the vector magnetic potential due to the current element and also the magnetic field intensity at the point (ρ, φ, z).

The vector potential in the differential form is given by:

$$dA=\frac{\mu_0 IdL}{4\pi R}$$

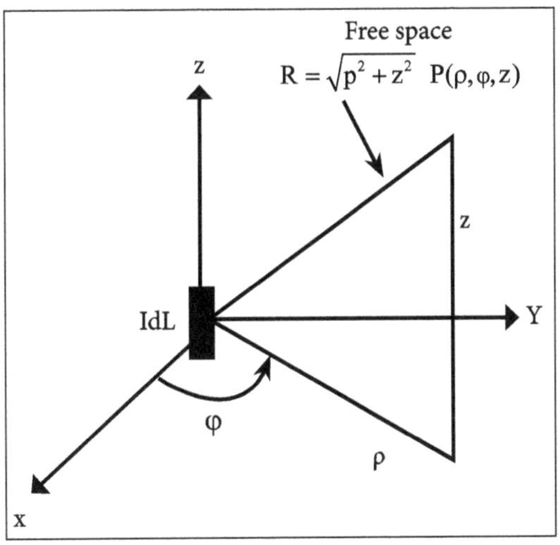

Let us consider a differential element Idz located at the origin in free space.

Consider a point P(ρ, φ, z) at which vector magnetic potential has to be determined.

$$dA=\frac{\mu_0 Idza_z}{4\pi\sqrt{\rho^2+z^2}}$$

The magnetic field intensity at P(ρ, φ, z) is given by:

$$dH=\frac{1}{\mu_0}\nabla\times dA=\frac{1}{\mu_0}\left(-\frac{\partial dA_z}{d\rho}\right)a_\phi$$

$$= \frac{Idz}{4\pi} \cdot \frac{\rho}{\left(\rho^2 + z^2\right)^{3/2}} a_\phi .$$

Magnetic field intensity along the axis of the circular coil:

Magnetic field intensity at point P due to current element Idl is given by,

$$dH = \frac{Idl \sin 90°}{4\pi r^2}$$

dH has two components dHy and dHz. Y component vanishes and only Z component exists.

$$dHz = dH \sin\alpha$$

$$= \frac{Idl}{4\pi r^2} \sin\alpha$$

$$Hz = \frac{I \sin\alpha}{4\pi r^2} \oint dl.$$

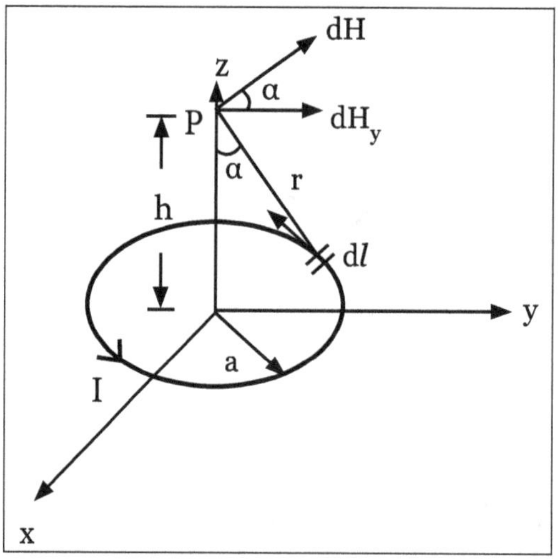

ϕ dl = Circumference of the loop = $2\pi a$

$$\therefore Hz = \frac{Ia \sin\alpha}{2r^2}$$

From the figure:

$$\sin\alpha = \frac{a}{r} \text{ and } r = \sqrt{h^2 + a^2}$$

$$\therefore \ Hz = \frac{Ia^2}{2\left(h^2 + a^2\right)^{3/2}}.$$

Magnetic field intensity along the center of the square loop:

Let us consider a square loop with side 'w' carrying a direct current I lies in the xy plane. Let O be the center of the loop at which the flux density has to be determined.

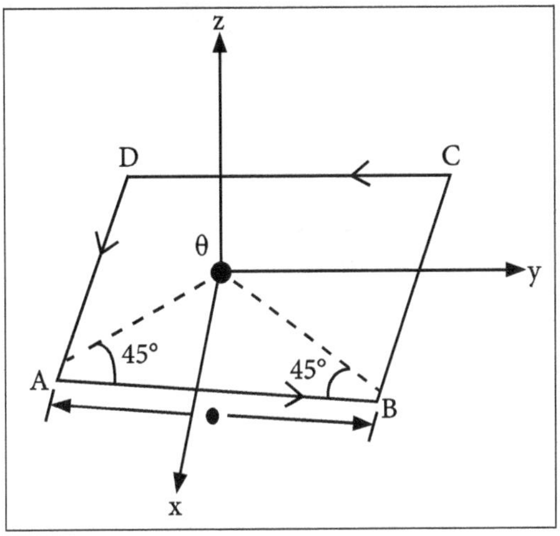

The field intensity at a distance h from any conductor of finite length is given by:

$$H = \frac{I}{4\pi h}\left[\cos\alpha_1 + \cos\alpha_2\right]$$

Let us consider side AB of the loop:

$$H_{AB} = \frac{I}{4\pi.R}\left[\cos 45^\circ + \cos 45^\circ\right]\hat{a}_z$$

$$= \frac{I}{4\pi R}\left[\frac{2}{\sqrt{2}}\right]\hat{a}_z$$

$$= \frac{I}{2\sqrt{2}\pi w}\hat{a}_z$$

The total field intensity at 'O' due to square loop ABCD.

$$= 4 \times \frac{I}{2\sqrt{2}\pi w}\hat{a}_z$$

$$H = \frac{\sqrt{2}I}{\pi w}\hat{a}_z$$

Magnetic field intensity due to infinite length current element:

Consider a straight conductor of length L with current I flowing through it. Consider a point P at a height h from this conductor in xy plane. Consider any current element Idx at a distance x from o. Let its distance to P be 'r'.

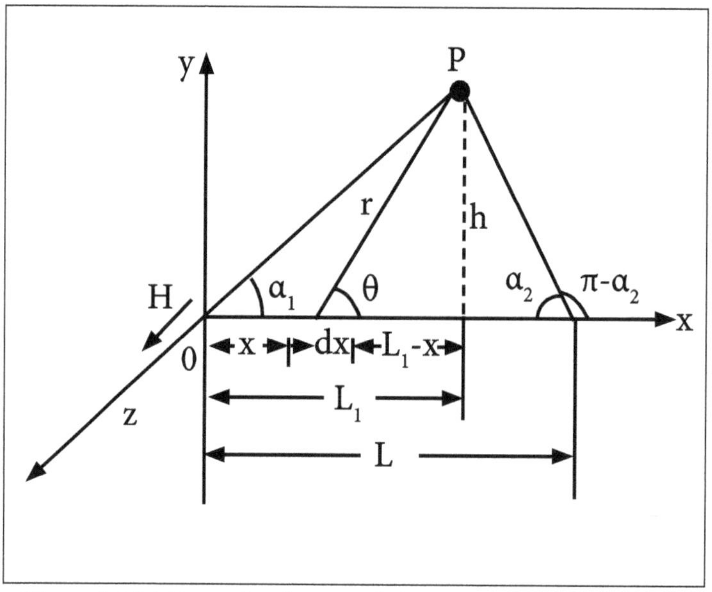

Magnetic Field Intensity due to Infinite Length Current Element.

Magnetic field intensity at P due to current element Idx is given by:

$$dH = \frac{I\,dx \sin\theta}{4\pi r^2} a_z$$

Magnetic field intensity due to the entire conductor is given by:

$$= \int \frac{I\,dx \sin\theta}{4\pi r^2} a_z \qquad ...(1)$$

From the figure:

$$\tan\theta = h/(L_1 - x)$$

$$L_1 - x = h\cot\theta$$

Differentiating:

$$-dx = h\left[-\text{cosec}^2\theta\, d\theta\right]$$

$$dx = h\,\text{cosec}^2\theta\, d\theta \qquad ...(2)$$

Again, referring with the above figure:

$$\sin\theta = \frac{h}{r}$$

$$r = \frac{h}{\sin\theta} = h\ \mathrm{cosec}\,\theta \qquad ...(3)$$

Substituting the values of dx and r from equations (2) and (3) in equation (1), we get:

$$\therefore \int dH = \int_{\alpha_1}^{\pi-\alpha_2} \frac{I \times h\ \mathrm{cosec}^2 d\theta \times \sin\theta}{4\pi\left[h^2\ \mathrm{cosec}^2\,\theta\right]} a_z$$

$$H = \frac{I}{4\pi h}\int_{\alpha_1}^{\pi-\alpha_2} \sin\theta\, d\theta\, a_z$$

$$H = \frac{I}{4\pi h}\left[-\cos\theta\right]_{\alpha_1}^{\pi-\alpha_2} a_z$$

$$H = \frac{I}{4\pi h}\left[-\cos(\pi-\alpha_2)+\cos\alpha_1\right] a_z$$

$$H = \frac{I}{4\pi h}\left[\cos\alpha_1 + \cos\alpha_2\right] a_z$$

For infinitely long conductor, $\alpha_1 = \alpha_2 = 0$.

$$\therefore\ H = \frac{I}{4\pi h}(2) a_z$$

$$H = \frac{I}{2\pi h} a_z.$$

Magnetic field intensity on the centre of the triangle:

Let us consider the side AB. The magnetic field intensity at 'o' due to AB is given by:

$$H = \frac{I}{4\pi h}\left[\cos\alpha_1 + \cos\alpha_2\right]\hat{a}_z$$

Here, $\alpha_1 = \alpha_2 = 30°$.

The height $CD = \sqrt{l^2 - (l/2)^2}$

$$= \sqrt{\frac{3l^2}{4}} = \sqrt{3}\frac{l}{2}.$$

The lengths $OD(h) = \frac{1}{3}CD = \frac{l}{2\sqrt{3}}$

So, $H_{AB} = \frac{I}{4\pi \times \frac{l}{2\sqrt{3}}}[\cos 30 + \cos 30]\hat{a}_z$

$$= \frac{\sqrt{3}\,I}{2\pi l} = \left(\frac{\sqrt{3}}{2} + \frac{\sqrt{3}}{2}\right) = \frac{3I}{2\pi l}\hat{a}_z$$

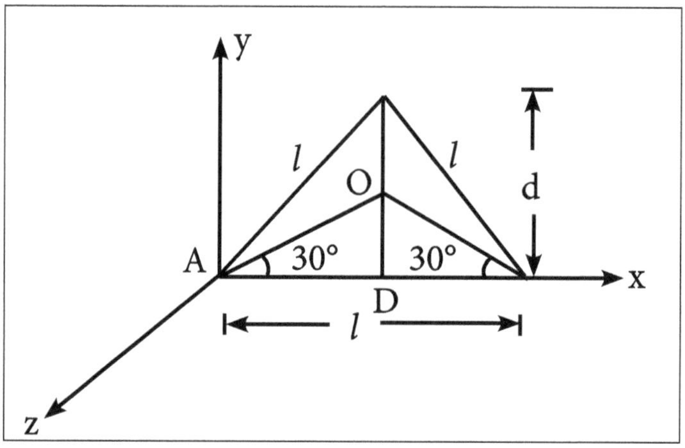

Magnetic field intensity on the centre of the triangle.

The contributions from sides BC and AC are also the same. Thus, the total magnetic field intensity at the center of the triangle is given by:

$$H = 3 \times \frac{3l}{2\pi l}\hat{a}_z = \frac{9l}{2\pi l}\hat{a}_z.$$

Problems

1. A circular loop located on $x^2 + y^2 = 4$, $z = 0$ carries direct current of 7 A along $\hat{a}_\phi$. Lets calculate the magnetic field intensity at (0, 0, – 5).

Solution:

Given:

$a = 2$, $I = 7$ A, $h = 5$

Formula to be used:

$$H = \frac{Ia^2}{2(h^2 + a^2)^{3/2}}.$$

$$= \frac{7 \times 2^2}{2\left(5^2 + 2^2\right)^{3/2}}$$

$$H = 0.0896\, A/m(-a_z)$$

2. If electric field intensity in free space is given by $\bar{E} = \frac{50}{\rho}\cos\left(10^8 t - 10z\right)\hat{a}_\rho v/m$. Let us calculate the magnetic field intensity.

Solution:

Given:

$$\bar{E} = \frac{50}{\rho}\cos\left(10^8 t - 10z\right)\hat{a}_\rho v/m$$

Formula to be used:

$$\nabla \times E = -\frac{\partial B}{\partial t}$$

$$-\frac{\partial B}{\partial t} = \frac{1}{\rho}\begin{vmatrix} a_\rho & \rho a_\phi & a_z \\ \frac{\partial}{\partial \rho} & \frac{\partial}{\partial \phi} & \frac{\partial}{\partial z} \\ \frac{50}{\rho}\cos\left(10^8 t - 10z\right) & 0 & 0 \end{vmatrix}$$

$$= -a_\phi\left[-\frac{\partial}{\partial z}\left(\frac{50}{\rho}\cos\left(10^8 t - 10z\right)\right)\right] + a_z[0]$$

$$= \frac{50}{\rho}\left(-\sin\left(10^8 t - 10z\right)\right)(-10)a_\phi$$

$$\frac{\partial B}{\partial t} = \frac{500}{\rho}\sin\left(10^8 t - 10z\right)a_\phi$$

$$B = \frac{500}{\rho}\int \sin\left(10^8 t - 10z\right)a_\phi dt$$

$$= \frac{500}{\rho}\left(-\cos\left(10^8 t - 10z\right)a_\phi\left(\frac{1}{10^8}\right)\right)$$

$$B = -\frac{5\times10^{-6}}{\rho}\cos\left(10^{8}t - 10z\right)a_{\phi}$$

$$H = \frac{B}{\mu o} = -\frac{5\times10^{-6}}{\rho\left(4\pi\times10^{-7}\right)}\cos\left(10^{8}t - 10z\right)a_{\phi}$$

$$= -\frac{3.98}{\rho}\cos\left(10^{8}t - 10z\right)a_{\phi}.$$

2.2 Magnetic Circuits

A magnetic circuit is made up of one or more closed loop paths containing a magnetic flux. The flux is usually generated by permanent magnets or electromagnets and confined to the path by magnetic cores consisting of ferromagnetic materials like iron, although there may be air gaps or other materials in the path.

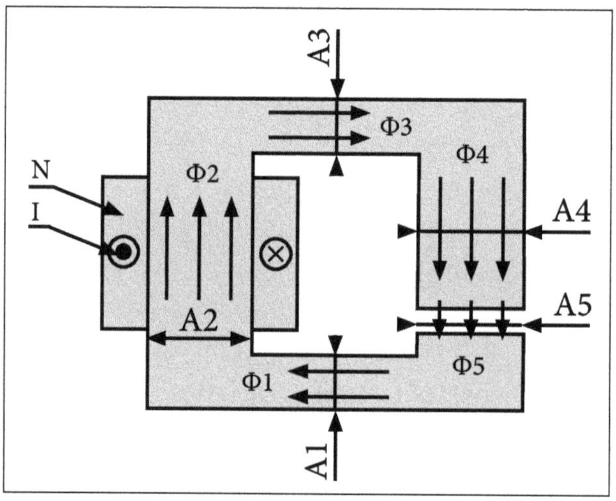

Magnetic circuit.

Magnetic Potential Energy

The energy of the magnetic moment 'm' in the externally produced magnetic field B, which has the potential energy,

$$U = -m.B$$

The magnetization M in a field is given by:

$$U = -\frac{1}{2}\int M.BdV$$

The magnet can also exert the non-contact force on the objects made up of steel, iron, cobalt or nickel. Either of the poles of magnet exerts the pulling force on these magnetic materials. The strength of the force increases as the magnet and the magnetic material are brought together.

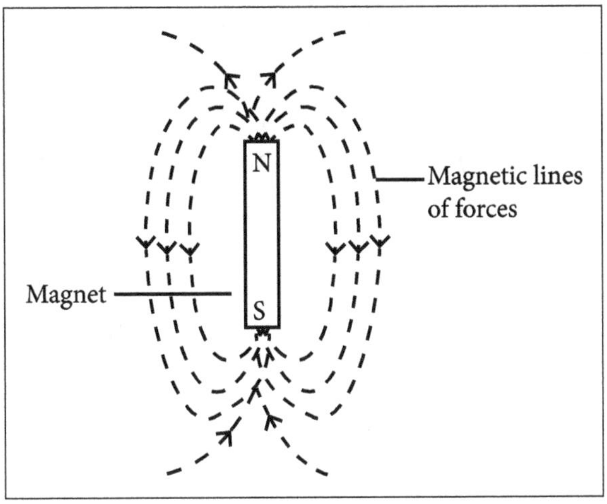

Magnet

2.2.1 Analogous Quantities in Magnetic and Electric Circuits

Analogy Between Electric and Magnetic Circuits

Magnetic circuits are analogous to the electric circuits:

- The common examples of the magnetic circuits are transformers, relays, generators and magnetic recording devices.
- An electric circuit forms a circuit (i.e. closed path) through which current flow. Similarly in magnetic circuits, magnetic lines of flux are continuous and can form closed paths.
- A single magnetic line of flux or all parallel magnetic lines of flux may be considered as magnetic circuits.
- Electromotive force in an electric circuit, in magnetic circuit is called magneto motive force (m.m.f.).
- The magneto motive force (m.m.f.) is defined as:

  $e_m = NI = \oint H.dL$.

The SI unit of m.m.f. is ampere (A):

- In electric circuit, resistance is defined as the ratio of voltage to current given by, R=V/I.

- In magnetic circuit, reluctance is the ratio of the magneto motive force to the total flux.

Similarities between magnetic circuit and electric circuit:

- Magnetic circuit follows equation (4) that is Ni = (Φ) (l / μA) or m.m.f (magneto motive force) = (Flux) (reluctance).
- Electric circuit follows ohm's law that is E = I.R or e.m.f (electro motive force) = (current) (Resistance). From above point m.m.f in magnetic circuit is like as e.m.f in electrical circuit.
- Flux in magnetic circuit is similar as current in electrical circuit.
- Reluctance in magnetic circuit, S = (l / μA) is similar to resistance R = (ρl/A) in electric circuit.
- Permeance (= 1/reluctance) in magnetic circuit is equivalent to conductance (=1/resistance) in electric circuit.

Differences between magnetic circuit and electric circuit:

- In magnetic circuit flux establishes but not flow like as current in magnetic circuit.
- In magnetic circuit energy needed only to establish the flux but no consistent energy need to maintain it whereas in electric circuit continuous energy needed to flow of current.
- Resistance of an electric circuit is constant (for same temperature) and is independent of current but reluctance of magnetic circuit is not constant because it depends on μ (=B/H) which is not constant and depends on B/H.

Uniqueness between electrical and magnetic circuits are given below:

S. No.	Electric Circuit	Magnetic Circuit
1.	EMF is the driving force in the electric circuit. Its unit is Volts.	MMF is the driving force in the magnetic circuit. Its unit is ampere turns.
2.	Path traced by the current is known as electric current.	Path traced by the magnetic flux is called as magnetic circuit.
3.	The flow of electrons decides the current in conductor.	The number of magnetic lines of force decides the flux.
4.	Resistance (R) oppose the flow of the current. Its unit is Ohm.	Reluctance (S) is opposed by magnetic path to the flux. Its unit is ampere turn/weber.
5.	There is a current I in the electric circuit which is measured in amperes.	There is flux φ in the magnetic circuit which is measured in the weber.

6.	$R = \rho.\, l/a$	$S = l/(\mu_0\mu_r a)$.
7.	Kirchhoff current law and voltage law is applicable to the electric circuit.	Kirchhoff MMF law and flux law is applicable to the magnetic flux.
8.	Current I = EMF/Resistance	Flux = MMF/Reluctance

2.2.2 Faraday's Law

First Law

Any change in the magnetic field of a coil of wire will cause an emf to be induced in the coil. This emf induced is called induced emf and if the conductor circuit is closed, the current will also circulate through the circuit and this current is called induced current.

Method to change magnetic field:

- By moving a magnet towards or away from the coil.
- By moving the coil into or out of the magnetic field.
- By rotating the coil relative to the magnet.
- By changing the area of the coil placed in the magnetic field.

Second Law

It states that the magnitude of emf induced in the coil is equal to the rate of change of flux that linkages with the coil. The flux linkage of the coil is the product of the number of turns in the coil and the flux associated with the coil.

Faraday's law states that the total emf induced in a circuit is equal to the rate of change of total magnetic flux linking the circuit.

Electric field produced by the changing magnetic field can be written as:

$$\text{emf} = -\frac{d\phi}{dt}\text{volts}$$

Faraday's Experiment

In this experiment, Faraday takes a magnet and a coil and connects a galvanometer across the coil. At starting, the magnet is at rest so there is no deflection in the galvanometer i.e., needle of galvanometer is at the center or zero position.

When the magnet is moved towards the coil, the needle of the galvanometer deflects in one direction. When the magnet is held stationary at that position, the needle of galvanometer returns back to zero position. Now when the magnet is moved away from the coil, there is some deflection in the needle, but in the opposite direction and again when the magnet becomes stationary, at that point with respect to coil, the needle of the galvanometer returns back to the zero position.

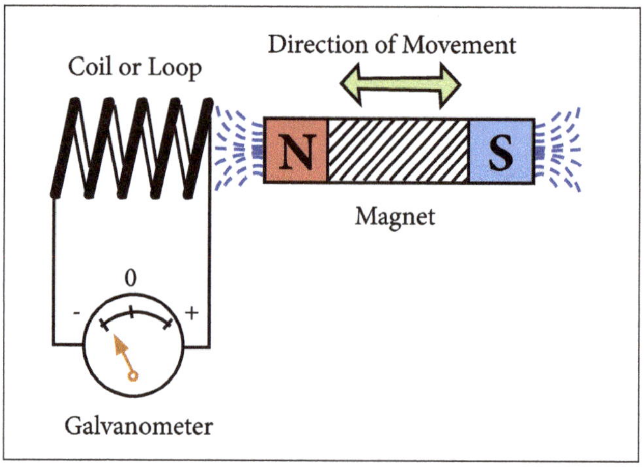

Similarly, if the magnet is held stationary and the coil is moved away and towards the magnet, the galvanometer shows deflection in the similar manner. It is also seen that, the faster the change in the magnetic field, the greater will be the induced emf or voltage in the coil.

Position of magnet	Deflection in galvanometer
Magnet at rest.	No deflection in galvanometer.
Magnet moves towards the coil.	Deflection in galvanometer in one direction.
Magnet is held stationary at the same position (i.e. near the coil).	No deflection in the galvanometer.
Magnet moves away from the coil.	Deflection in galvanometer, but in opposite direction.
Magnet is held stationary at same position (away from the coil).	No deflection in galvanometer.

Let's consider a magnet approaching towards a coil. Here, we consider 2 instants at time T_1 and time T_2.

Flux linkage with the coil at time, $T_1 = N\varphi_1$ Wb.

Flux linkage with the coil at time, $T_2 = N\varphi_2$ wb.

Change in flux linkage = $N(\varphi_2 - \varphi_1)$.

Let this change in flux linkage be, $\varphi = \varphi_2 - \varphi_1$.

So, the change in flux linkage = $N\varphi$.

Now the rate of change of flux linkage = $N\varphi/t$.

Take derivative on right hand side we will get.

The rate of change of flux linkage = N dφ/dt.

But according to Faraday's laws of electromagnetic induction, the rate of change of flux linkage is equal to the induced emf.

$$E = N\frac{d\phi}{dt}$$

Considering Lenz's Law

$$E = -N\frac{d\phi}{dt}$$

Where, flux φ in Wb = B.A

B = Magnetic field strength.

A = Area of the coil.

Applications of Faraday Law

Faraday law is one of the basic and important laws of electromagnetism. This law finds its application in most of the electrical machines, industries and medical field etc.

- Electrical Generators: The basic working principle of electrical generator is Faraday's law of mutual induction. Electric generator is used to convert mechanical energy into electrical energy.
- Electrical Transformers: It is a static AC device which is used to either step up or step down voltage or current. It is used in generating station, transmission and distribution system. The transformer works on Faraday's law.
- Electromagnetic Flow Meters: It is used to measure velocity of blood and certain fluids. When a magnetic field is applied to electrically insulated pipe in which conducting fluids are flowing, then according to Faraday's law, an electromotive force is induced in it. This induced emf is proportional to velocity of fluid flowing.
- Induction Cookers: The Induction cooker, is a most fastest way of cooking. It also works on principle of mutual induction. When current flows through the coil of copper wire placed below a cooking container, it produces a changing magnetic field. This alternating or changing magnetic field induces an emf and hence the current in the conductive container and we know that flow of current always produces heat in it.
- Musical Instruments: It is also used in musical instruments like electric guitar, electric violin etc.

- Form the bases of Electromagnetic Theory: Faraday's idea of lines of force is used in well-known Maxwell's equations. According to Faraday's law, change in magnetic field gives rise to change in electric field and the converse of this is used in Maxwell's equations.

2.3 Self and Mutual Inductance

Self-inductance of a coil is defined as the ratio of self-induced emf to the rate of change of current in the coil. It is denoted by 'L' and it depends upon the physical characteristics of the coil. Unit of self-inductance is Henry.

Consider a coil consists of N turns and carrying current I in the counter clockwise direction, as shown in figure. If the current is steady, then the magnetic flux through the loop will remain constant. Suppose the current I changes with time, then according to Faraday's law, an induced emf will arise to oppose the change.

The induced current will flow clockwise if $dI/dt > 0$ and counter clockwise if $dI/dt < 0$. The property of the loop in which its own magnetic field opposes any change in current is termed as "self-inductance" and the emf generated is called the self-induced emf or back emf. All current-carrying loops exhibit this property. In particular, an inductor is a circuit element which has a large self-inductance.

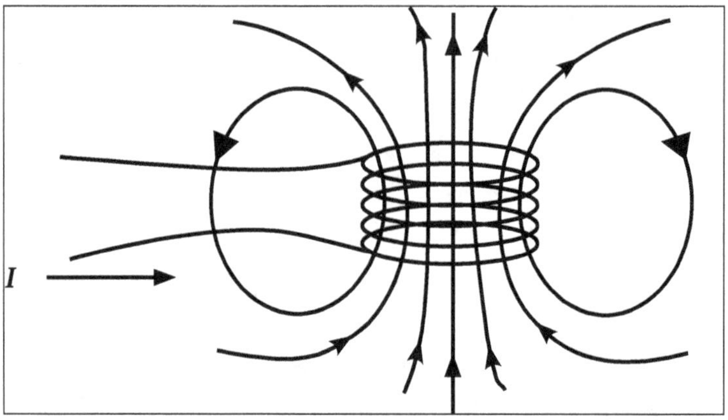

Magnetic flux through the current loop.

Mathematically, the self-induced emf can be written as:

$$\varepsilon_L = -N\frac{d\Phi_B}{dt} = -N\frac{d}{dt}\iint \vec{B}.d\vec{A}$$

And is related to the self-inductance L by:

$$\varepsilon_L = -L\frac{dI}{dt}$$

The two expressions can be combined to yield:

$$L = \frac{N\Phi_B}{I}$$

Physically, the inductance L is a measure of an inductor's resistance to the change of current. Larger the value of L, lower the rate of change of current.

Mutual Inductance

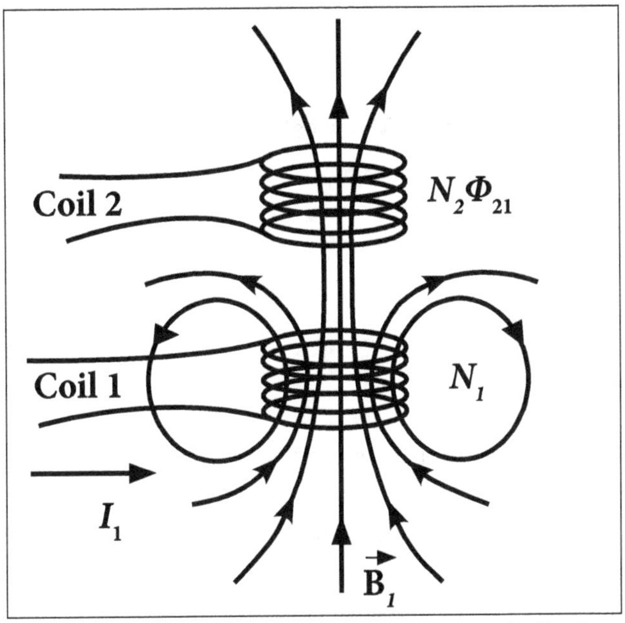

Changing current in coil 1 produces changing magnetic flux in coil 2.

Mutual inductance is a property associated with two or more coils or inductors which are in close presence of common magnetic flux which links the coils. A transformer is such a device whose operation is based on mutual inductance.

Let us consider two coils L_1 & L_2 which are close together, so that the flux produced by i_1 in coil L_1 also link coil L_2. When a voltage V_1 is applied across L_1, a current I_1 will start flowing through L_1, produce a flux ϕ. This flux ϕ also links coil L_2. If the current flowing through L_1 changes with respect to time, then the induced flux ϕ also changes with respect to time.

The time varying flux surrounding the second coil L_2, induces an electromagnetic force (emf) or voltage across the coil L_2. This voltage is proportional to the rate of change of current flowing through coil L_1. Then, the two coils are said to be inductively coupled and is called mutual inductance.

Suppose two coils are placed near each other, as shown in the figure. The first coil has N_1 turns and carries a current I_1 which gives rise to a magnetic field B_1. Since the two coils are close to each other, some of the magnetic field lines through coil 1 will also pass through coil 2. Let ϕ_{21} denote the magnetic flux through one turn of coil 2 due to I_1. Now, by varying I_1 with time, there will be an induced emf associated with the changing magnetic flux in the second coil.

$$\text{i.e., } \varepsilon_{21} = -N_2 \frac{d\Phi_{21}}{dt} = -\frac{d}{dt} \iint_{\text{coil}\,2} \vec{B}_1 . d\vec{A}_2$$

The time rate of change of magnetic flux ϕ_{21} in coil 2 is proportional to the time rate of change of the current in coil 1.

$$\text{i.e., } N_2 \frac{d\Phi_{21}}{dt} = M_{21} \frac{dI_1}{dt}$$

Where, the proportionality constant M_{21} is called the mutual inductance. It can also be written as:

$$M_{21} = \frac{N_2 \Phi_{21}}{I_1}$$

The SI unit for inductance is henry (H).

$$1 \text{ henry} = 1\text{ H} = 1\text{ T m}^2/\text{A}$$

The mutual inductance M_{21} depends only on the geometrical properties of the two coils such as the number of turns and the radii of the two coils. In a similar manner, instead there is a current I_2 in the second coil and it is varying with time, then the induced emf in coil 1 becomes:

$$\varepsilon_{12} = -N_1 \frac{d\Phi_{12}}{dt} = -\frac{d}{dt} \iint_{\text{coil}\,1} \vec{B}_2 . d\vec{A}_1 \quad \text{and a current is induced in coil 1.}$$

This changing flux in coil 1 is proportional to the changing current in coil 2:

$$N_1 \frac{d\Phi_{12}}{dt} = M_{12} \frac{dI_2}{dt}$$

Where, the proportionality constant M_{12} is another mutual inductance and can be written as:

$$M_{12} = \frac{N_1 \Phi_{12}}{I_2}$$

However, using the reciprocity theorem which combines the Ampere's law and the Biot-Savart law, we may show that the constants are equal as given below:

$$M_{12} = M_{21} \equiv M$$

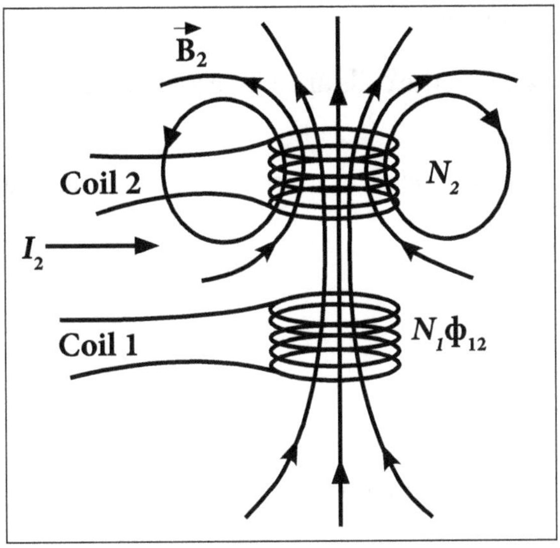

Changing current in coil 2 produces changing magnetic flux in coil 1.

Coefficient of Coupling

In coupled coils, the coefficient of coupling is defined as the fraction of the total flux produced by one coil linking another coil.

The amount of coupling between the inductively coupled coils is expressed in terms of the coefficient of coupling, which is defined as:

Coefficient of coupling:

$$K = \frac{\varphi_{12}}{\varphi_1} = \frac{\varphi_{21}}{\varphi_2}$$

[At K=1, maximum mutual inductance occurs in coupled circuits].

Where,

M - Mutual inductance between the two coils.

L_1 - Self-inductance of the first coil.

L_2 - Self-inductance of the second coil.

Problem

1. An iron ring of relative permeability 100 is wound uniformly with two coils of 100 and 400 turns of wire. The cross-section of the ring is $4cm^2$. The mean circumference is 50cm. Let's calculate:

- The self-inductance of each of the two coils.

- The mutual inductance.
- Total inductance when the coils are connected in series with the flux in the same sense.
- Total inductance when the coils are connected in series with flux in the opposite sense.

Solution:

Given:

Relative permeability = 100

Number of turns of wire = 100 and 400

Cross-section of the ring = 4cm^2

Mean circumference = 50cm

Formula to be used:

$$L = \frac{\mu_0 N^2 A}{2\pi R_m}$$

$$M = \frac{\mu_0 N_1 N_2 A}{2\pi R_m}$$

Total inductance = $L_1 + L_2 + 2M$

- Self-inductance of coil with 100 turns is given by:

$$L = \frac{\mu_0 N^2 A}{2\pi R_m} = \frac{4\pi \times 10^{-7} \times 100^2 \times 4}{50 \times 10^{-2}}$$

$= 0.1$ Henry.

Self-inductance with 400 turns is given by:

$$L = \frac{4\pi \times 10^{-7} \times 400^2 \times 4}{50 \times 10^{-2}} = 1.61 \text{Henry}$$

- Mutual inductance of the coil is given by:

$$M = \frac{\mu_0 N_1 N_2 A}{2\pi R_m}$$

$$= \frac{4\pi\times10^{-7}\times100\times400\times4}{50\times10^{-2}}$$

$= 0.402$ Henry.

- Total inductance:

$$= L_1 + L_2 + 2M$$
$$= 0.1 + 1.61 + 2\times0.4$$
$$= 2.51.$$

- Total inductance when the coils are in series with flux in the opposite sense is given by:

$$= L_1 + L_2 - 2M$$
$$= 0.1 + 1.61 - 2\times0.4$$
$$= 0.91.$$

2.3.1 Energy Stored in a Magnetic Field

An inductor in the circuit serves to oppose any change in the current through it, work must be done by an external source such as the battery to establish the current in the inductor. From the work-energy theorem, we conclude that the energy can be stored in the inductor.

The power or rate at which an external emf works to overcome the self-induced emf and pass current I in the inductor is given by:

$$P_L = \frac{dW_{ext}}{dt} = -I\varepsilon_{ext}$$

If only the external emf and the inductor are present, then $\varepsilon ext = -\varepsilon L$:

$$P_L = \frac{dW_{ext}}{dt} = -I\varepsilon_L = +IL\frac{dI}{dt}$$

The total work done by the external source to increase the current form zero to I is then:

$$W_{ext} = \int dW_{ext} = \int_0^I LI'dI' = \frac{1}{2}LI^2$$

This is equal to the magnetic energy stored in the inductor:

$$U_B = \frac{1}{2}LI^2$$

The above expression is analogous to the electric energy stored in the capacitor,

$$U_E = \frac{1}{2}\frac{Q^2}{C}$$

Problems

1. A coil is wound on an iron core to form a solenoid. A certain current is passed through the coil which is producing a flux of 40 μ Wb. The length of magnetic circuit is 75 cm while its cross-sectional area is 3 cm². Let us calculate the energy stored in the circuit. Assume relative permeability of iron as 1500.

Solution:

Given:

$$l = 75 \text{ cm} = 0.75 \text{ m},\ a = 3 \text{ cm}^2 = 3\times10^{-4} \text{ m}^2$$

$$\varphi = 40\ \mu\text{Wb} = 40\times10^{-6} \text{ Wb},\ \mu r = 1500$$

Formula to be used:

$$B = \frac{\phi}{a}$$

$$E/\text{unit volume} = \frac{1}{2}\frac{B^2}{\mu}\text{Joules/m}^3$$

$$B = \frac{\phi}{a} = \frac{40\times10^{-6}}{3\times10^{-4}} = 0.133\,\text{Wb/m}^2$$

Energy stored per unit volume:

$$\frac{1}{2}\frac{B^2}{\mu} = \frac{1}{2}\frac{B^2}{\mu_0\mu_r} = \frac{1}{2}\frac{(0.133)^2}{4\pi\times10^{-7}\times1500}$$

$$= 4.7157\ \text{J/m}^3$$

Total energy stored = energy per unit volume × volume = $E\times(a\times l)$

$$= 4.7157\times(3\times10-4\times0.75)$$

$$= 0.00106 \text{ Joules.}$$

2. Let us find the inductance of a coil of 200 turns wound on a paper core tube of 25 cm length and 5 cm radius. Also calculate energy stored in it if current rises from zero to 5 A.

Solution:

Given:

N = 200, 1 = 25 cm = 0.25 m, r = 5 cm = 0.05 m

Formula to be used:

$$S = \frac{1}{\mu_0 a}$$

$$c/s \text{ area} = \frac{\pi}{4} d^2$$

$$L = \frac{N^2}{S}$$

$$E = \frac{1}{2} LI^2$$

$$c/s \text{ area} = \frac{\pi}{4} d^2$$

Where, d = diameter

$$a = \frac{\pi}{4}(2r)^2 = \frac{\pi}{4} \times (2 \times 0.05)^2 = 7.853 \times 10^{-3} m^2$$

For paper $\mu_r = 1$

$$S = \frac{1}{\mu_0 a} = \frac{0.25}{4\pi \times 10^{-7} \times 7.853 \times 10^{-3}} = 2.533 \times 10^7 \ AT/Wb$$

$$L = \frac{N^2}{S} = \frac{(200)^2}{2533 \times 10^7} = 1579 \times 10^{-3} \quad H = 1.579 \ mH$$

$$E = \frac{1}{2} LI^2 = \frac{1}{2} \times 1579 \times 10^{-3} \times (5)^2 = 0.01973 \ J.$$

2.3.2 B-H Curve

There are two methods by which B-H curve can be obtained for the magnetic material specimen.

Method of Reversals

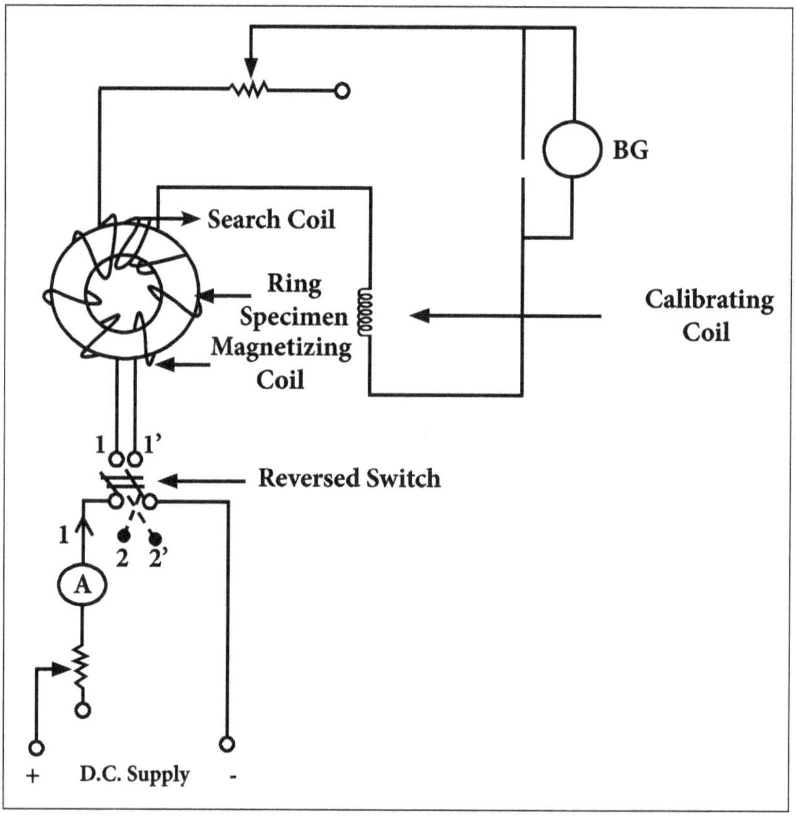

Method of Reversal - circuit diagram.

A ring specimen with known dimensions is taken for the test. A thin tape is wound on the ring. The search coil insulated by the paraffined wax is wound over the tape. Another layer of tape is wound on the search coil and then the magnetizing winding is wound uniformly on the specimen. The overall circuit used is as shown in the figure.

The complete specimen is demagnetized before the test. Using the variable resistance, the magnetizing current is adjusted to its lower value, at the beginning of the test. The ballistic galvanometer switch K is closed and reversing switch S is thrown backward and forward for about twenty times. This brings the iron specimen into a reproducible cyclic magnetic state.

The galvanometer key K is now opened and the flux in the specimen corresponding to this value of H is measured from the deflection of the ballistic galvanometer, by reversing the switch S. The change in flux is measured by the galvanometer, when the reversing switch S is quickly reversed. It will be twice the flux in the specimen, corresponding to the value of H applied. This value of H can be obtained as:

$$H = \frac{NI_1}{l}$$

Where,

N = Number of turns on the magnetizing winding.

I_1 = Corresponding magnetizing current.

l = Mean circumference length of specimen in m.

While the flux density B is obtained by dividing the flux measured by the area of cross-section of the specimen. The procedure is repeated for different values of H by increasing H up to the maximum testing point value. The graph of B against H gives the required B-H curve for the specimen.

Step by Step Method

In this method, reversal of magnetizing current is not used. The magnetizing current in the winding is supplied through a potential divider as shown in the figure. The potential divider has number of tappings. The tappings are arranged in such a way that the magnetizing force H increases in suitable number of steps, up to the required maximum value.

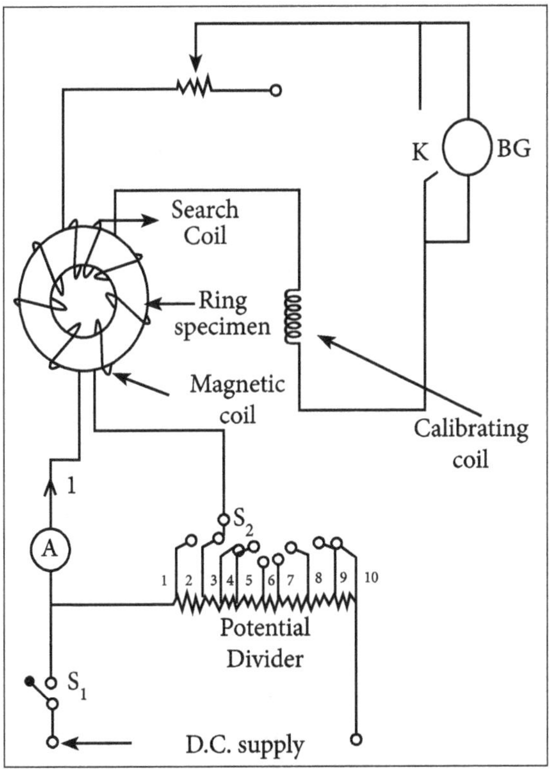

Step by step method - Circuit diagram.

The specimen is completely demagnetized before starting the test. The switch S_1 is closed with switch S_2 at tapping 1. Due to this, there will be some change in the flux. Hence, there will be increase in the flux density from 0 to B_1. This value can be obtained by observing the deflection of the ballistic galvanometer and recording the value of corresponding magnetizing current. The corresponding value of the magnetizing force H_1 can be obtained.

The switch S_2 is instantaneously changed to tapping 2 which increases the magnetizing force to H_2. Due to this there is increase in flux and hence flux density by the amount ΔB. This can be determined from the galvanometer throw. Hence, B_2 at H_2 can be obtained as $B_1 + \Delta B$.

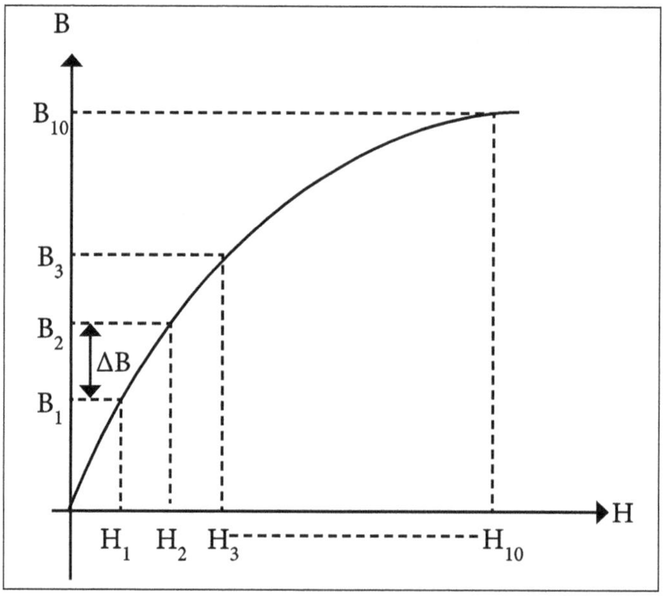

B-H curve from step by step method.

The procedure is repeated for various tappings till maximum value of H is achieved. The graph of B against H is then plotted which is nothing but the B-H curve for the specimen under test.

2.4 Hysteresis and Eddy Current Losses

Hysteresis Loss

It happens in the armature winding of the DC machine since any given part of the armature is exposed to magnetic field of reversal as it passes underneath sequence poles. The below figure shows the 2-pole DC machine of rotating armature.

Consider a tiny low piece ab of the armature winding. Once the piece ab is underneath N-pole, the magnetic lines pass from a to b. Half a revolution well along, identical piece of iron is underneath S-pole and magnetic lines pass from b to a in order that magnetism within the iron is overturned.

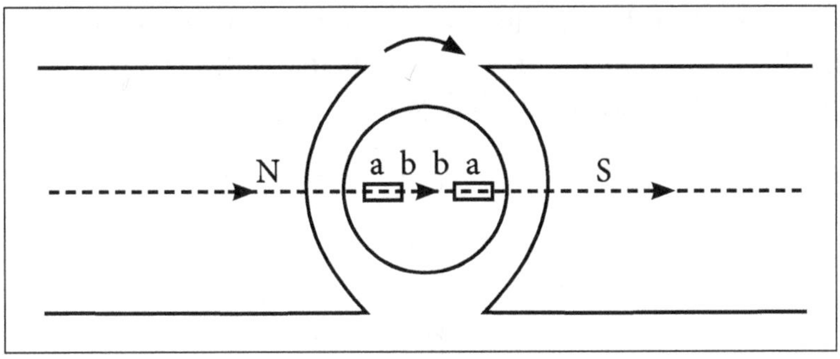

Two pole DC machine.

So, as to reverse constantly the molecular magnets within the armature core, particular quantity of power must be spent that is named hysteresis loss. It is given by Steinmetz formula.

Hysteresis loss, $P_h = \eta B^{16}{}_{max} fV$ watts.

Where,

η = Steinmetz hysteresis co-efficient.

B_{max} = Maximum flux Density in armature winding.

F = Frequency of magnetic reversals.

= NP/120 (N is in RPM).

V = Volume of armature in m^3.

Eddy Current Loss

A voltage is induced in a conductor when it is situated in a changing magnetic field. To produce strong magnetic effects, coils of electromagnets are wound on iron cores. When the current in the coil of an electromagnet is changing, the core constitutes a conductor in a changing magnetic field.

Voltages are induced in the core and circulating currents are caused to flow. These circulating currents are called eddy currents. The flow of these currents causes losses as a waste of energy and heating of core.

In DC circuits, eddy current loss is not serious, since an induced voltage is produced only when the circuit is switched ON or OFF. In an AC circuits, the flux changes continuously and eddy current loss becomes appreciable.

It has been found that eddy currents travel crosswise in the cores. The magnitude of eddy currents can be reduced to a very small value by increasing the cross-sectional resistance of the iron.

The cores are made of thin sheets called laminations. Each lamination is about 0.5 mm thick and is insulated from the others by a coating of oxides or varnish or both, so that it offers a high resistance to the flow of eddy currents.

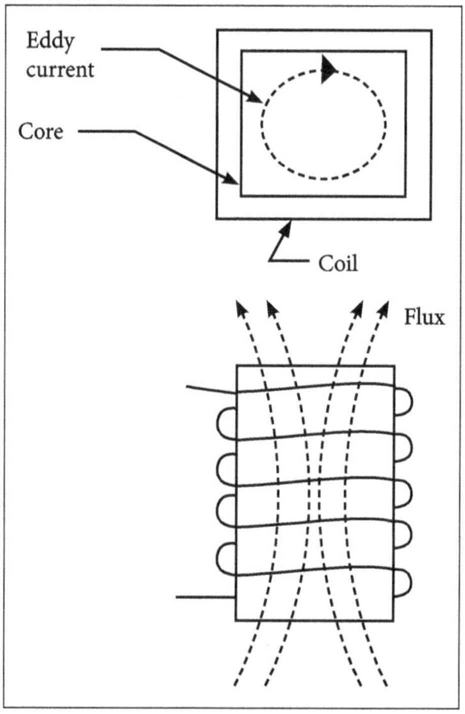

Flow of eddy currents in iron core.

The use of laminated cores is to reduce the active cross-sectional area by a small amount (about 5-10%) due to thickness of insulation.

Eddy current loss is given by:

$$P_e = k_e f^2 B_m^{\ 2} t^2 \left(\text{volume}\right)$$

Where,

P_e = Eddy current loss in Watts.

f = Frequency, Hz.

B_m = Maximum flux density, T.

Problems

1. Consider the total core loss of a specimen of silicon steel is found to be 1500 W at 50 Hz. Keeping the flux density constant the loss becomes 3000 W when the frequency is raised to 75 Hz. Let us calculate the hysteresis and eddy current loss at each of the frequencies separately.

Solution:

Given:

Total core loss of a specimen of silicon steel - 1500 W at 50 Hz,

Flux density - Constant

Loss - 3000 W when the frequency is raised to 75 Hz.

For constant flux density, total core loss can be expressed as:

$$P = Af + Bf^2 \text{ (or) } P/f = A + Bf$$

$$1500/50 = A + 50B \text{ (or) } 30 = A + 50B \quad \text{...(1)}$$

$$3000/75 = A + 75B \text{ (or) } 40 = A + 75B \quad \text{...(2)}$$

Solving equation (1) and (2), we get A = 10, B = 2/5.

Therefore, $P = 10f + 2/5f^2 = P_n + P_e$

At 50Hz,

$$P_h = 10 \times 50 = 500 \text{ W}$$

$$P_e = 2/5 \times 2500 = 1000 \text{ W}.$$

At 75 Hz,

$$P_h = 10 \times 75 = 750 \text{ W}$$

$$P_e = 2/5 \times 75^2 = 2250 \text{ W}.$$

2. Consider the core-loss (hysteresis + eddy-current loss) for a given specimen of magnetic material is found to be 2000 W at 50 Hz. Keeping the flux density constant, the frequency of the supply is raised to 75 Hz resulting in a core loss of 3200 W, let us compute the hysteresis and eddy current losses at both the frequencies separately.

Solution:

Given:

Core-loss of magnetic material - 2000 W at 50 Hz.

Frequency of the supply is raised to 75 Hz in a core loss of 3200 W.

Since, the flux density is the same in both cases.

We can use the relation:

Total core loss $Wi = A_f + B_f^2$

$$\text{or } W_i / f = A + B_f$$

$$\therefore \frac{2000}{50} = A + 50B \Rightarrow 40 = A + 50B \quad ...(1)$$

$$\frac{3200}{75} = A + 75B \Rightarrow 42.67 = A + 75B \quad ...(2)$$

Solving (1) and (2):

(1)	⇒	40	=	A + 50	B
(2)	⇒	42.67	=	A + 75	B
	(–)			(–)	
	–	2.67	=	A – 25	B
	⇒	B = 0.11			

Substitute $B = 0.11$ in $(1) \Rightarrow 40 = A + 50\ B$

$$\Rightarrow 40 = A + 50(0.11)$$

$$A = 34.5.$$

Therefore, A = 34.5 and B = 0.11.

At 50 Hz, the two losses are hysteresis loss:

$$W_n = A_f = 34.5 \times 50 = 1725\ W = 1.725\ kW \text{ and}$$

Eddy current loss, $W_e = B_f^2 = 0.11 \times 50^2 = 275$ W.

2.4.1 Lifting Power of Electromagnet

An electromagnet is often used to pull or lift iron pieces or objects made of iron. This is possible due to the energy stored in the magnetic field of the electromagnet.

In the below figure an electromagnet having N number of turns and the coil carrying a current, I is shown. The cross-sectional area of the core is A. An iron piece is separated from the electromagnet by a distance, l.

Energy stored in a magnetic field is given by:

$$W = \frac{1}{2} LI^2 J$$

Putting,

$$L = N\frac{\phi}{I}$$

We get,

$$W = \frac{1}{2} N\phi IJ$$

Again,

$$H = \frac{NI}{\ell} \text{ and } \varphi = B \times A$$

Therefore,

$$W = \frac{1}{2} NI\phi = \frac{1}{2} H \times \ell \times B \times A$$

$$= \frac{1}{2} HBA\ell J$$

Again,

$$B = \mu_o H$$

Or,

$$H = \frac{B}{\mu_o}$$

Considering the magnetic field in the air gap between the magnet and the piece of iron, energy stored is given by,

$$W = \frac{1}{2}\frac{B}{\mu_o} BA\ell J$$

$$= \frac{B^2}{2\mu_o}(A \times \ell)J \quad ...(1)$$

(A × L) is the volume of the air gap between the electromagnet and the iron piece.

Force, F = Work done per unit length

$= (W/l)$

Or,

$$F = \frac{B^2 A}{2\mu_0} N \qquad ...(2)$$

The stored energy of the magnetic field between the electromagnet and the piece of iron is able to pull the iron piece towards it. When the exciting coil of the electromagnet is energized by passing current, the force of attraction will pull the iron near to the face of the electromagnet. Such electromagnets can be used to pull or lift large amounts of magnetic material and shift the material from one place to another. Thus, the lifting power of the electromagnet can be used to do some mechanical work. For example, electromagnets are used to lift iron ores from the place of storage and bring them for processing when required.

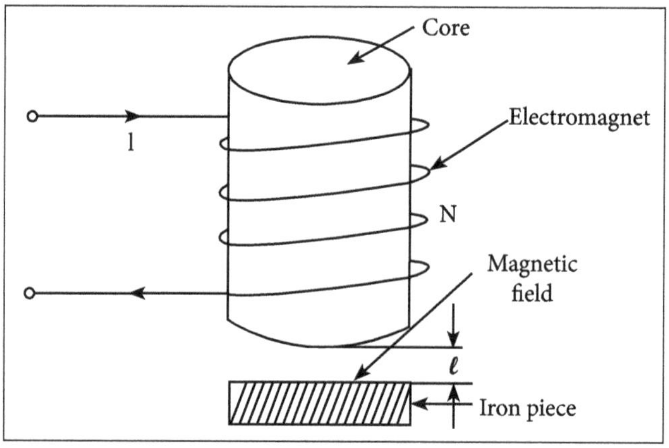

Lifting power of an electromagnet.

3

AC Fundamental

3.1 Production of Alternating Voltage, Waveforms, Average and RMS Values

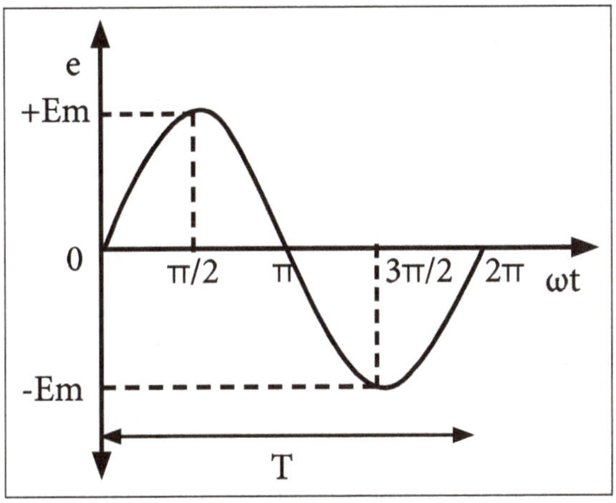

An alternating quantity changes continuously in magnitude and alternates in direction at regular intervals of time. Important terms associated with an alternating quantity are defined below:

- Amplitude: It is the maximum value attained by an alternating quantity. Also called as maximum or peak value.
- Time Period (T): It is the Time Taken in seconds to complete one cycle of an alternating quantity.
- Instantaneous Value: It is the value of the quantity at any instant.
- Frequency (f): It is the number of cycles that occur in one second. The unit of frequency is Hz or cycles/sec.

The relationship between time period and frequency can be derived as follows:

Time taken to complete f cycles = 1 second.

Time taken to complete 1 cycle = 1/f second.

$T = 1/f$

Advantages of AC System Over DC System

- AC voltages can be efficiently stepped up or down using transformer.
- Switchgear for AC system is simpler than DC system.
- AC motors are cheaper and simpler in construction than DC motors.

Generation of sinusoidal AC voltage:

Consider a rectangular coil of N turns placed in a uniform magnetic field as shown in the figure. The coil is rotating in the counter clockwise direction at an uniform angular velocity of ω rad/sec.

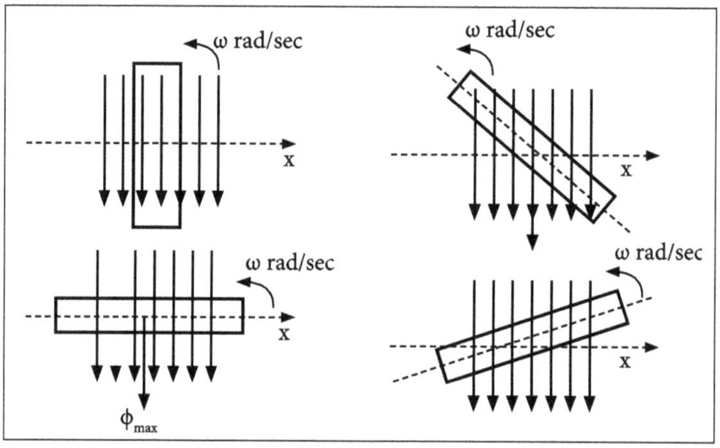

When the coil is in the vertical position, the flux linking the coil is zero because the plane of the coil is parallel to the direction of the magnetic field. Hence at this position, emf induced in the coil is zero. When the coil moves by some angle in the anticlockwise direction, there is a rate of change of flux linking the coil and hence an emf is induced in the coil.

When the coil reaches the horizontal position, flux linking the coil is maximum and hence, the emf induced is also maximum. When the coil further moves in the anticlockwise direction, the emf induced in the coil reduces. Next when the coil comes to the vertical position, the emf induced becomes zero.

After that the same cycle repeats and the emf is induced in the opposite direction. When the coil completes one complete revolution, one cycle of AC voltage is generated.

The generation of sinusoidal AC voltage can also be explained using mathematical equations.

Consider a rectangular coil of N turns placed in a uniform magnetic field in the position shown in the figure. The maximum flux linking the coil is in the downward direction as shown in the figure.

This flux can be divided into two components, one component acting along the plane of the coil φ_{max} sinωt and another component acting perpendicular to the plane of the coil φ_{max} cosωt.

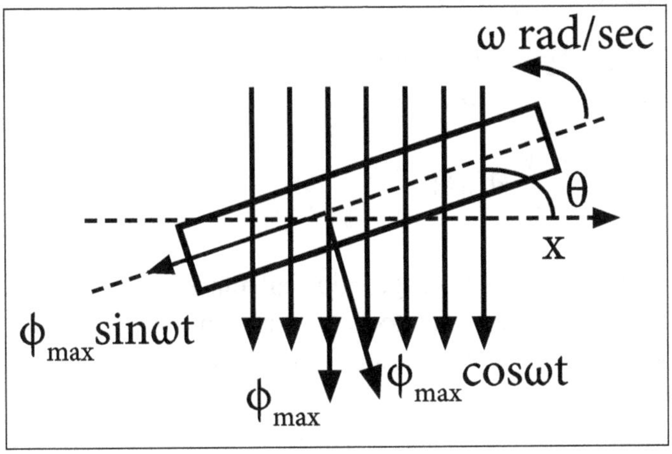

The component of flux acting along the plane of the coil does not induce any flux in the coil. Only the component acting perpendicular to the plane of the coil, (i.e.) φ_{max} cosωt induces an emf in the coil.

$$\phi = \phi_{max} \cos\omega t$$

$$e = -N\frac{d\Phi}{dt}$$

$$e = -N\frac{d}{dt}\Phi_{max}\cos\omega t$$

$$e = N\phi_{max}\omega \sin\omega t$$

$$e = E_m \sin\omega t$$

Hence, the emf induced in the coil is a sinusoidal emf. This will induce a sinusoidal current in the circuit given by:

$$i = I_m \sin \omega t$$

Angular Frequency (ω)

Angular frequency is defined as the number of radians covered in one second ((i.e.) the angle covered by the rotating coil). The unit of angular frequency is rad/sec.

$$\omega = \frac{2\pi}{T} = 2\pi f$$

Average Value

The arithmetic average of all the values of an alternating quantity over one cycle is called its average value.

$$\text{Average value} = \frac{\text{Area under one cycle}}{\text{Base}}$$

$$V_{av} = \frac{1}{2\pi}\int_{0}^{2\pi} v\,d(\omega t)$$

For symmetrical waveforms, the average value calculated over one cycle becomes equal to zero because the positive area cancels the negative area. Hence for symmetrical waveforms, the average value is calculated for half cycle.

$$\text{Average value} = \frac{\text{Area under one half cycle}}{\text{Base}}$$

$$V_{av} = \frac{1}{\pi}\int_{0}^{\pi} v\,d(\omega t)$$

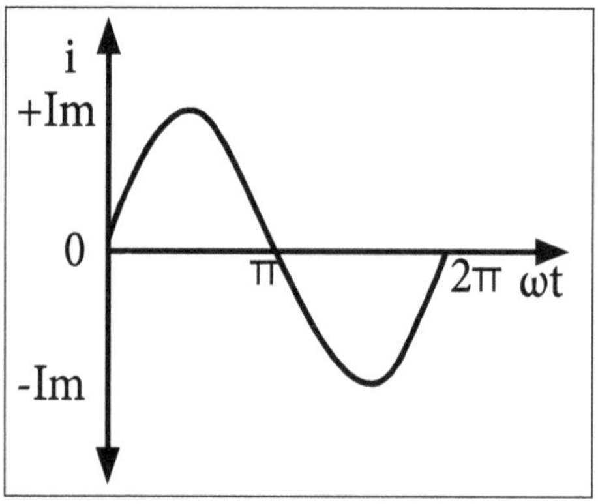

Average value of a sinusoidal current.

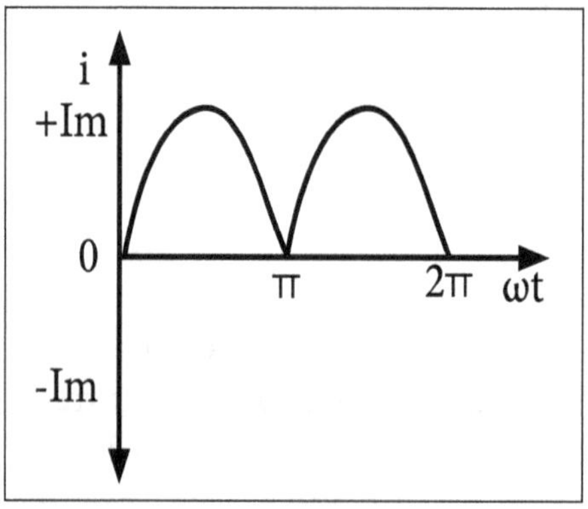

Average value of a full wave rectifier output.

$$i = I_m \sin \omega t$$

$$I_{av} = \frac{1}{\pi}\int_0^{\pi} i\,d(\omega t)$$

$$I_{av} = \frac{1}{\pi}\int_0^{\pi} I_m \sin \omega t\,d(\omega t)$$

$$I_{av} = \frac{2I_m}{\pi} = 0.637 I_m$$

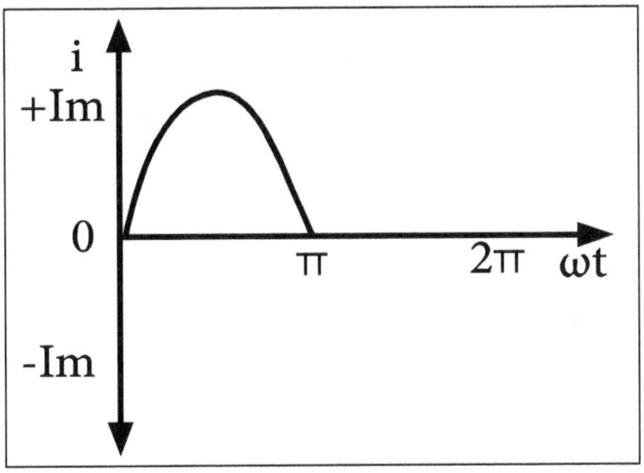

Average value of a half wave rectifier output.

$$i = I_m \sin \omega t$$

$$I_{av} = \frac{1}{2\pi}\int_0^{\pi} i\,d(\omega t)$$

$$I_{av} = \frac{1}{2\pi}\int_0^{\pi} I_m \sin \omega t\,d(\omega t)$$

$$I_{av} = \frac{I_m}{\pi} = 0.318\ I_m$$

RMS or Effective Value

The RMS or effective value of an alternating quantity is that steady current which when flowing through a given resistance for a given time produces the same amount of heat produced by the alternating current flowing through the same resistance for the same time.

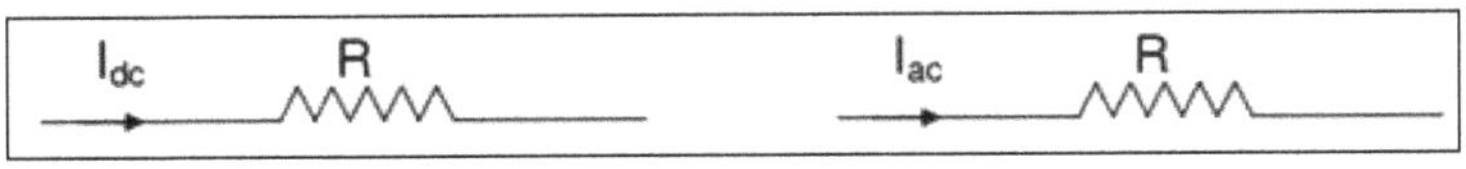

$$RMS = \sqrt{\frac{\text{Area under squared curve}}{\text{base}}}$$

$$V_{rms} = \sqrt{\frac{1}{2\pi}\int_{0}^{2\pi} v^2 d(\omega t)}$$

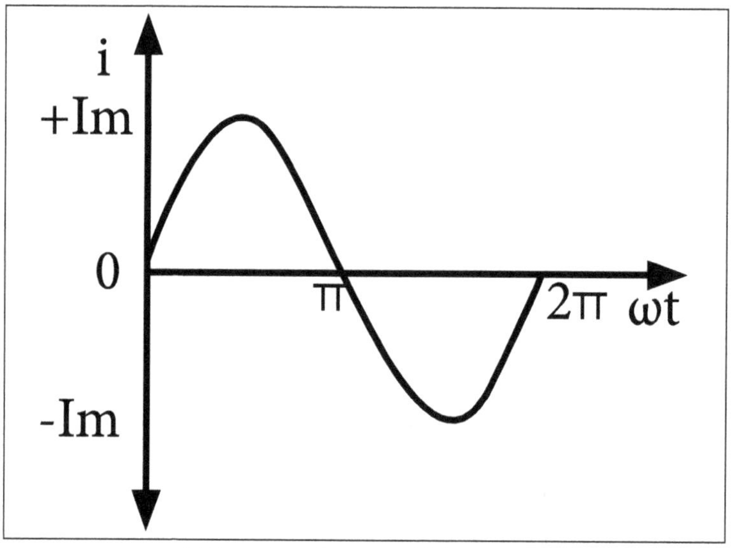

RMS value of a sinusoidal current.

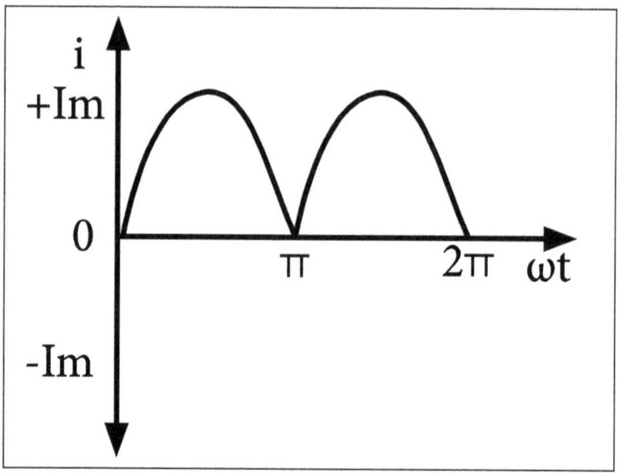

RMS value of a full wave rectifier output.

$$i = I_m \sin \omega t$$

$$I_{rms} = \sqrt{\frac{1}{2\pi}\int_{0}^{2\pi} i^2 d(\omega t)}$$

$$I_{rms} = \sqrt{\frac{1}{\pi}\int_{0}^{\pi} I_m^2 \sin^2 \omega t d(\omega t)}$$

$$I_{rms} = \frac{I_m}{\sqrt{2}} = 0.707 I_m$$

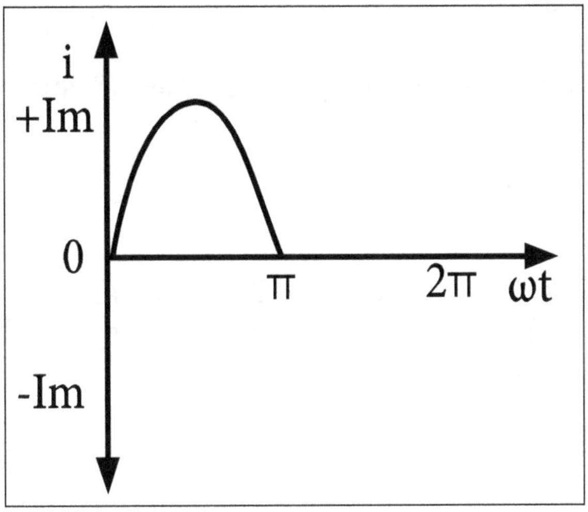

RMS value of a half wave rectifier output.

$$i = I_m \sin \omega t$$

$$I_{rms} = \sqrt{\frac{1}{2\pi}\int_0^{2\pi} i^2 d(\omega t)}$$

$$I_{rms} = \sqrt{\frac{1}{\pi}\int_0^{\pi} I_m^2 \sin^2 \omega t d(\omega t)}$$

$$I_{rms} = \frac{I_m}{2} = 0.5 I_m$$

3.1.1 Peak, Form Factor and Phase Factor

Peak Factor or Crest Factor

Crest Factor is the ratio between the peak value and the RMS value of the waveform and is given as:

$$\text{Crest Factor} = \frac{\text{peak value}}{\text{R.M.S value}} = \frac{V\max}{0.707 \times V\max}$$

As the peak value is always higher than the rms value, crest factor is always higher than one. For a sinusoidal alternating quantity the crest factor is given by:

$$\frac{\sqrt{2} \times \text{RMS value}}{\text{RMS value}}$$

$$K_p = \sqrt{2} \text{ or } 1.414$$

For a pure sinusoidal waveform the Crest Factor will always be equal to 1.414.

Form Factor

The form factor of an alternating quantity is defined as the ratio of its RMS value to its average value.

$$\text{Form factor } K_f = \frac{\text{RMS value}}{\text{Average value}} \qquad ...(1)$$

Form factor is a dimensionless quantity and its value is always higher than one. That means the RMS value will always be higher than the average value. Form factor of a sinusoidal alternating current is given by,

$$K_f = \frac{I_{rms}}{I_{av}} = \frac{0.707 I_m}{0.637 I_m} = 1.11$$

The value of $K_f = 1.11$ is valid only for the sinusoidal (sine or cosine) ac quantities. For all the other shapes of ac quantities the form factor is different from 1.11 and it can be obtained by using the general definition given in equation (1).

Phase

Phase is a measure of relative time difference between two sine waves. Even though phase is truly a time difference, it is almost always measured in terms of angle, either degrees or radians. This represents normalization to the time taken by one cycle of the wave in question, without regard to its true time period.

Phase Difference:

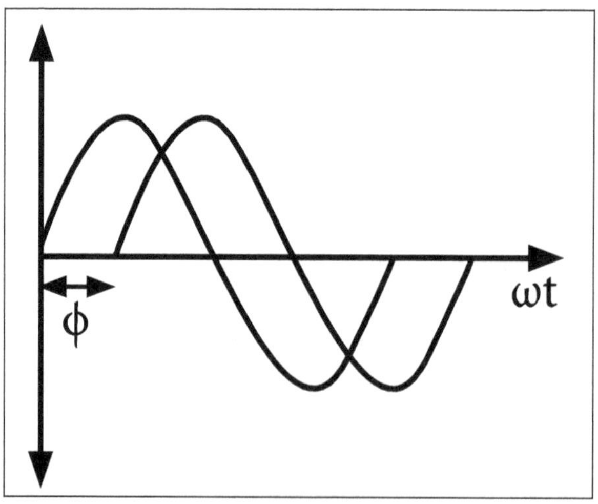

When two alternating quantities of the same frequency have different zero points, they are said to have a phase difference. The angle between the zero points is the angle of phase difference.

In Phase

Two waveforms are said to be in phase, when the phase difference between them is zero. That is the zero points of both the waveforms are same. The waveform, phasor and equation representation of two sinusoidal quantities which are in phase is as shown. The figure shows that the voltage and current are in phase.

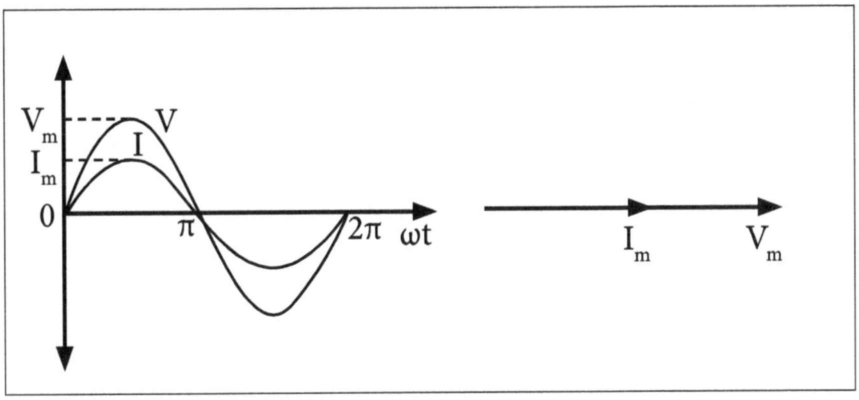

$$v = V_m \sin \omega t$$

$$i = I_m \sin \omega t$$

3.1.2 Phasor Representation and Phasor Diagram

Phasor Representation

We can represent an alternating quantity such as a sine or cosine wave with the help of a phasor. So "phasor" representation which is a very simple way to represent a sinusoidal quantity is being used.

Definition and graphical representation of a phasor:

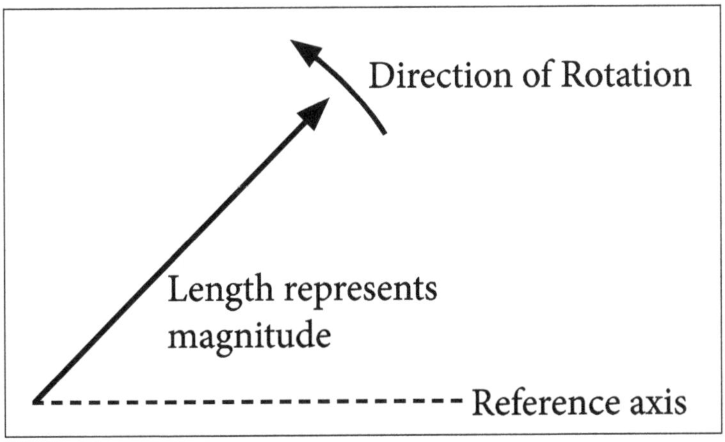

Phasor representation of sinusoidal quantities.

A phasor is a straight line with an arrow marked on one side. The length of this straight line represents the magnitude of the sinusoidal quantity being represented and the arrow represents its direction. Therefore, phasor representation is similar to the vector representation.

However the phasors rotate in the anticlockwise direction as shown in the below figure. The length of the phasor represents the rms value of the sinusoidal quantity. Sometimes, the length also represents the peak value.

Speed of rotation of the phasor is equal to ω radians/sec, where $\omega = 2\pi f$. One rotation of the phasor corresponds to one cycle of the alternating waveform as shown in the below figure:

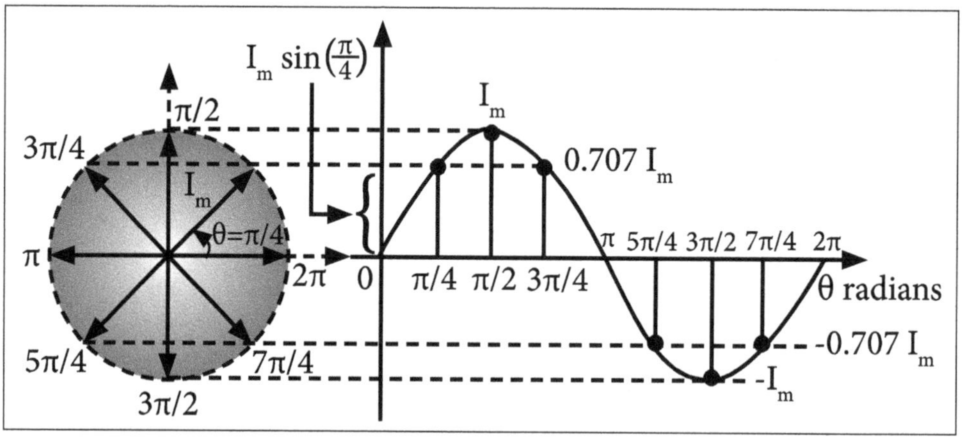

Relation between an alternating quantity and phasor.

If the length of the phasor is equal to the peak value of the sinusoidal AC quantity, its angular velocity $\omega = 2\pi f$ and it rotates in the anticlockwise direction in space than at any given angle θ, its projection on the Y-axis gives us the instantaneous value of the sinusoidal AC quantity at that angle.

We can represent two or more sinusoidal quantities simultaneously on the same phasor diagram if and only if their frequencies are same. If their frequencies are not same, then they cannot be represented on the same phasor diagram.

Phasor Diagram for Resistor

In case of resistor, the voltage and the current are in same phase or we can say that the phase angle difference between voltage and current is zero. According to Ohm's Law, the instantaneous voltage V across a resistor is proportional to the instantaneous current I flowing through it.

$$V = R.I$$

$$V = R.I_m \sin\omega t$$

$V = V_m \sin\omega t$

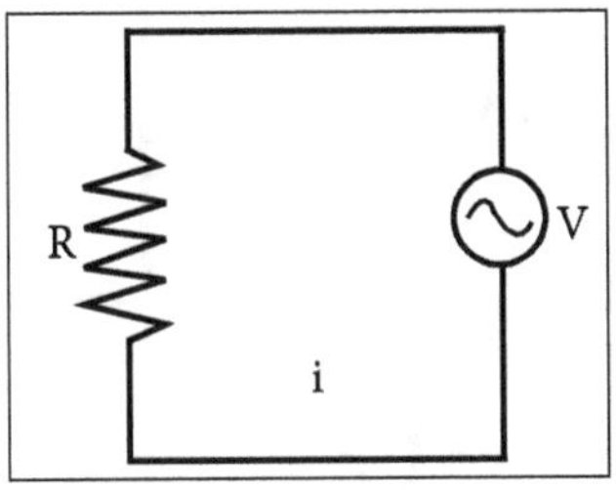

Resistor in an AC circuit.

The two graphs show the voltage across the resistor and the current flowing through the resistor as a function of time.

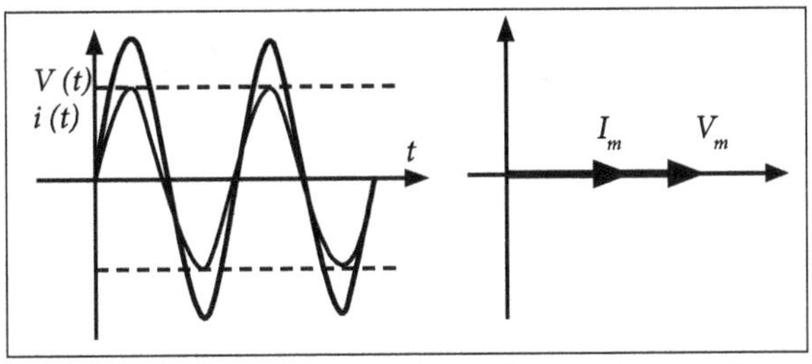

Sinusoidal variation in phase V and I & Phasor diagram for resistor.

Thus, voltage and current are in phase.

Phasor Diagram for Inductor

In inductor, the voltage and the current are not in phase. The voltage leads that of current by 90° or in the other words, voltage attains its maximum and zero value, 90° before the current attains it.

$V_m = X_L . I_m$

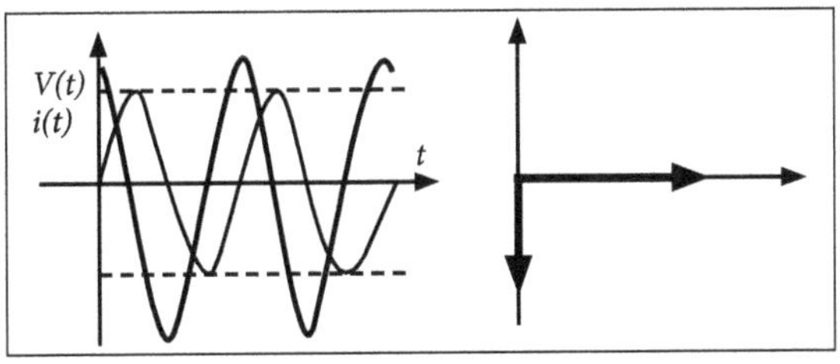

Sinusoidal variation in phase V and I & Phasor diagram for inductor.

Phasor Diagram for Capacitor

In case of capacitor, the current leads the voltage by 90° or in the other words, voltage attains its maximum and zero value 0° after the current attains it, i.e., the phasor diagram of capacitor is exactly opposite of inductor.

$$V_m = X_C . I_m$$

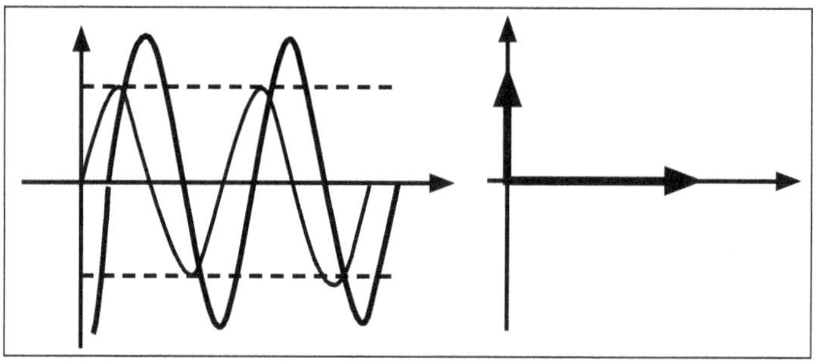

Sinusoidal variation in phase V and I & Phasor diagram for capacitor.

3.2 Behavior of AC Series and Parallel Circuits

Series RL Circuit

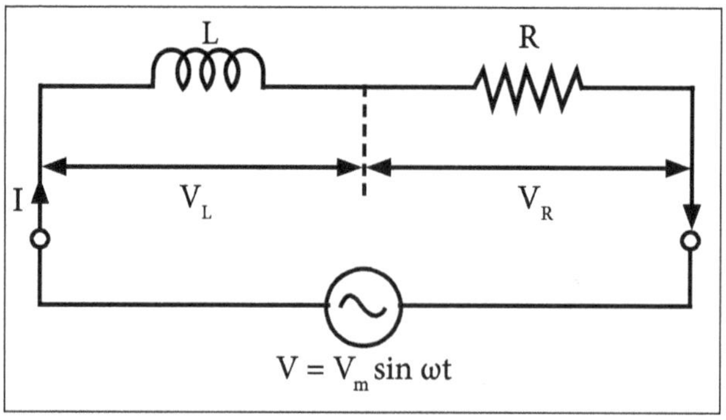

(a) Series RL circuit.

Consider a circuit consisting of pure resistance R ohms connected in series with a pure inductance of L henries as shown in the above figure (a). The series combination is connected across a.c. supply given by $v = V_m \sin \omega t$.

Circuit draws a current I then there are two voltage drops:

- Drop across pure resistance, $V_R = I \times R$.
- Drop across pure inductance, $V_L = I \times XL$.

Where,

$$X_L = 2\pi fL$$

I = rms value of current drawn.

V_R, V_L = rms values of voltage drops.

Kirchhoff's voltage law can be applied to a.c. circuit, but addition of voltages should be a phasor addition and not algebraic addition.

$$\overline{V} = \overline{V_R} + \overline{V_L}$$
$$\overline{V} = \overline{IR} + \overline{IX_L}$$

(Phasor addition).

Following are the steps to draw the phasor diagram:

- Take current as a reference phasor.
- In case of resistance, voltage and current are in phase, so VR will be along the current phasor.
- In case of inductance, current lags voltage by 90°, but, as current is the reference, VL must be shown leading with respect to the current by 90°.
- The supply voltage being the vector sum of these two vectors VL and VR obtained by law of parallelogram.

From the voltage triangle, we can write:

$$V = \sqrt{(V_R)^2 + (V_L)^2} = \sqrt{(IR)^2 + (I \times X_L)^2}$$

$$= I\sqrt{(R)^2 + (X_L)^2}$$

$$V = IZ$$

While,

$$Z = \sqrt{(R)^2 + (X_L)^2}$$

(Impedance of the circuit).

The impedance Z is measured in ohms.

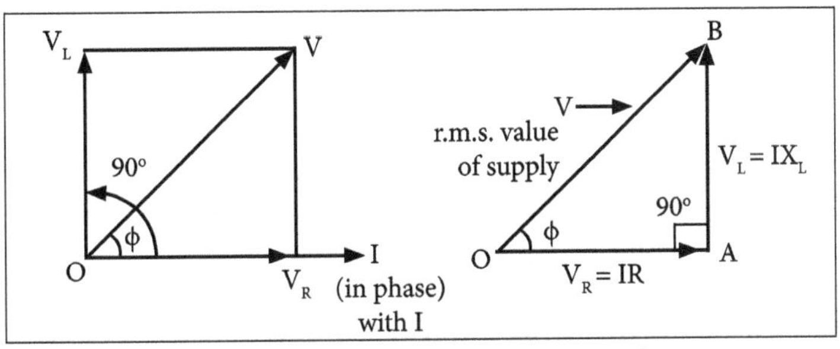

(b) Phasor diagram. (c) Voltage triangle.

RLC Series Circuit

Consider the driven series RLC circuit with $V(t)=V_o\sin(\omega t+\varphi)$ as shown in the below figure.

The current in the circuit is given by:

$$I(t)=I_o\sin\omega t \quad ...(1)$$

Notice that we have added a phase constant φ for V(t) and I(t) when we were analysing single element driven circuits. Applying Kirchhoff's modified loop rule, we obtain:

$$V(t)-V_R(t)-V_L(t)-V_C(t)=0 \quad ...(2)$$

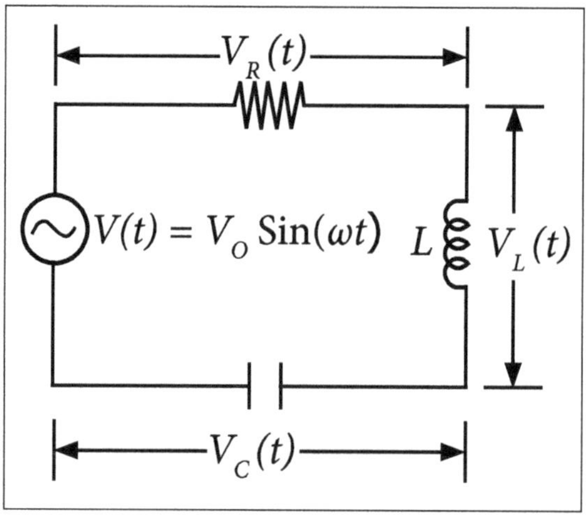

Driven series RLC Circuit.

We can rewrite equation (2) using $V_R(t)$ = IR, $V_L(t)$ = L dI/dt and $V_C(t)$ = Q/C as:

$$L=\frac{dI}{dt}+IR+\frac{Q}{C}=V_o\sin(\omega t+\phi) \quad ...(3)$$

Differentiate equation (3) using I = +dQ/dt and divide through by L, yields a second order damped linear driven differential equation:

$$\frac{d^2I}{dt^2}+\frac{R}{L}\frac{dI}{dt}+\frac{I}{LC}=\frac{\omega V_o}{L}\cos(\omega t+\phi) \quad ...(4)$$

We shall find the amplitude, I_0 of the current and phase constant φ which is the phase shift between the voltage source and the current by examining the phasors associates with the three circuit elements R, L and C.

The instantaneous voltages across each of the three circuit elements has different amplitude and phase compared to the current, as seen from the phasor diagrams.

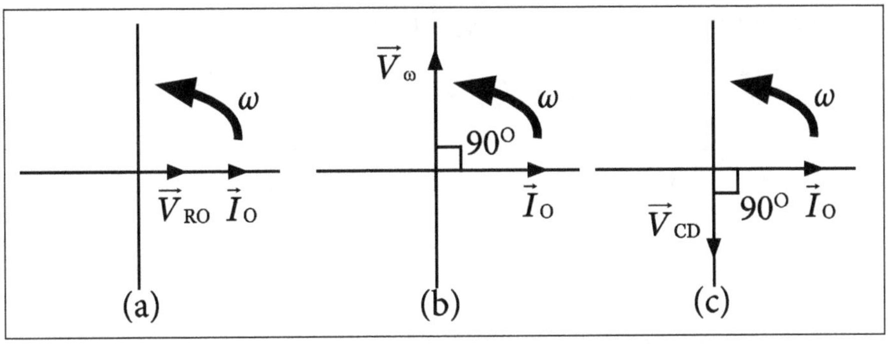

Phasor diagrams for the relationships between current and voltage in: (a) the resistor, (b) the inductor and (c) the capacitor, of a series RLC circuit.

Using the phasor representation, equation (2) can be written as:

$$\vec{V}_O = \vec{V}_{RO} + \vec{V}_{LO} + \vec{V}_{CO} \quad ...(5)$$

Again we can see that the current phasor Io leads the capacitive voltage phasor VCo by $\pi/2$, but lags the inductive voltage phasor V_{Lo} by $\pi/2$. The three voltage phasors rotate counter clockwise as time increases with their relative positions fixed.

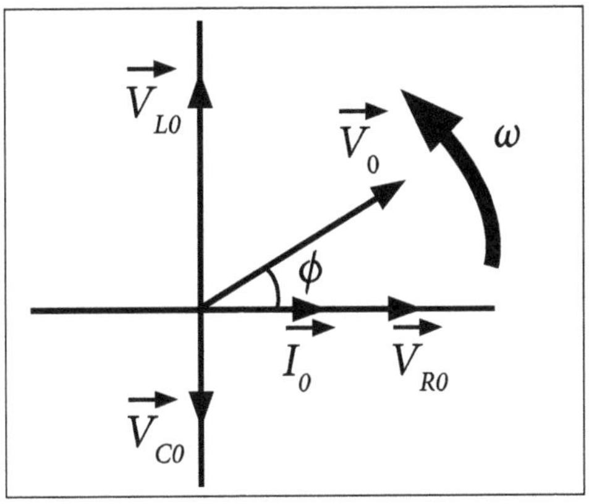

Phasor diagram for the series RLC circuit.

The relationship between different voltage amplitudes is depicted in the below figure:

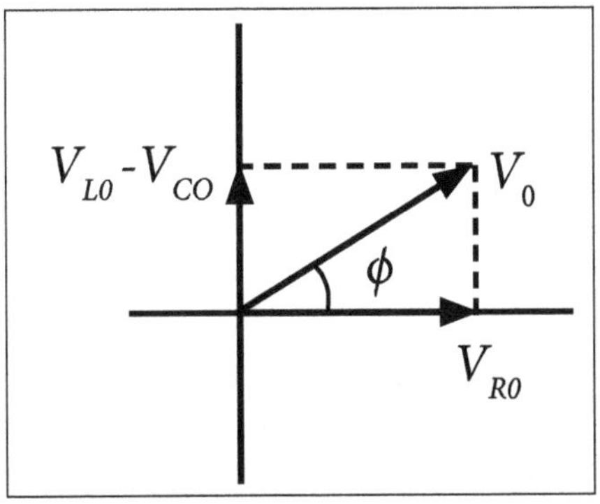

Voltage relationship.

From the above figure, we see that the amplitude satisfies:

$$V_o = \left|\vec{V}_o\right| = \left|\vec{V}_{Ro} + \vec{V}_{Lo} + \vec{V}_{Co}\right| = \sqrt{V_{Ro}^2 (V_{Lo} - V_{Co})^2}$$

$$= \sqrt{(I_o X_R)^2 + (I_o X_L - I_o X_C)^2}$$

$$= I_o \sqrt{X_R^2 + (X_L - X_C)^2} \qquad ...(6)$$

Therefore, the amplitude of the current is:

$$I_o = \frac{V_o}{\sqrt{X_R^2 + (X_L - X_C)^2}} \qquad ...(7)$$

Equation (7) becomes:

$$I_o = \frac{V_o}{\sqrt{R^2 + \left(\omega L - \dfrac{1}{\omega C}\right)^2}}, \text{ series RLC circuit} \qquad ...(8)$$

We can determine that the phase constant satisfies:

$$\tan\phi = \left(\frac{X_L - X_C}{X_R}\right) = \frac{1}{R}\left(\omega L - \frac{1}{\omega C}\right) \qquad ...(9)$$

Therefore, the phase constant is:

$$\phi = \tan^{-1}\frac{1}{R}\left(\omega L - \frac{1}{\omega C}\right), \text{ series RLC circuit} \qquad ...(10)$$

It is crucial to note that the maximum amplitude of the AC voltage source V_o is not equal to the sum of the maximum voltage amplitudes across the three circuit elements.

$$V_o \neq V_{Ro} + V_{Lo} + V_{Co}$$

This is due to the fact that the voltages are not in phase with one another and they reach their maxima at different times.

RL Parallel Circuit

In RL parallel circuit, resistor and inductor are connected in parallel with each other and this combination is supplied by a voltage source, V_{in}. The output voltage of circuit is V_{out}. Since the resistor and inductor are connected in parallel, the input voltage is equal to output voltage but the currents flowing in resistor and inductor are different.

The parallel RL circuit is not used as filter for voltages because in this circuit, the output voltage is equal to the input voltage and thus, it is not commonly used as compared to series RL circuit.

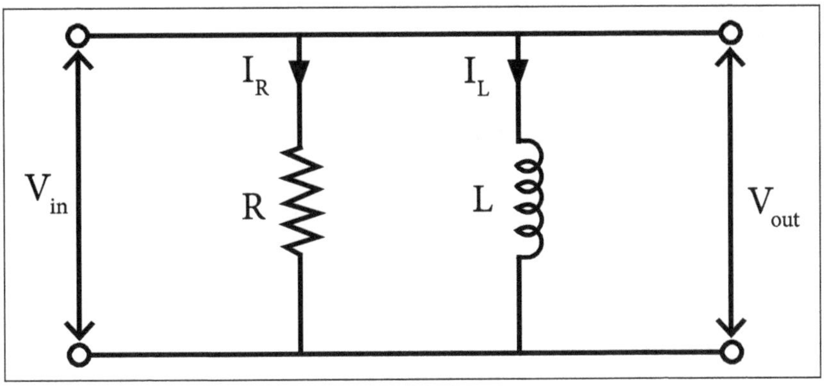

RL parallel circuit.

Let,

I_R = Current flowing in the resistor branch in amperes.

I_T = Total current flowing from voltage source in amperes.

θ = Angle between IR and IT.

I_L = Current flowing in the inductor branch in amperes.

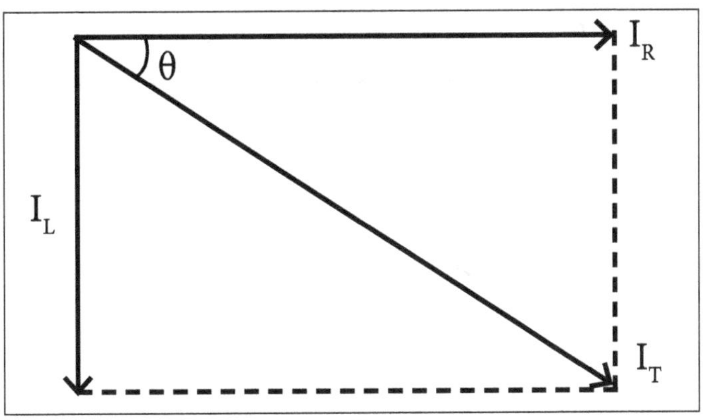

Phasor diagram.

So, the total current I_T is expressed as:

$$I_T^2 = I_R^2 + I_L^2$$

In complex form, the currents are written as:

$$I_R = \frac{V_{in}}{R}$$

$$I_L = \frac{V_{in}}{j\omega L}$$

$$\Rightarrow I_L = -j\frac{V_{in}}{\omega L}\left(\text{where}\frac{1}{j} = -j\right)$$

Therefore, total current can be written as:

$$I_T = \frac{V_{in}}{R - j\frac{V_{in}}{\omega L}}$$

RLC Parallel Circuit

Consider a RLC circuit in which resistor, inductor and capacitor are connected in parallel to each other. This parallel combination is supplied by a voltage supply, V_S. This parallel RLC circuit is exactly opposite to the series RLC circuit.

In the series RLC circuit, the current flowing through all the three components remains the same, but in parallel circuit, the voltage across each element remains the same and the current gets divided in each component depending upon the impedance of each component. Therefore, parallel RLC circuit is said to have dual relationship with series RLC circuit.

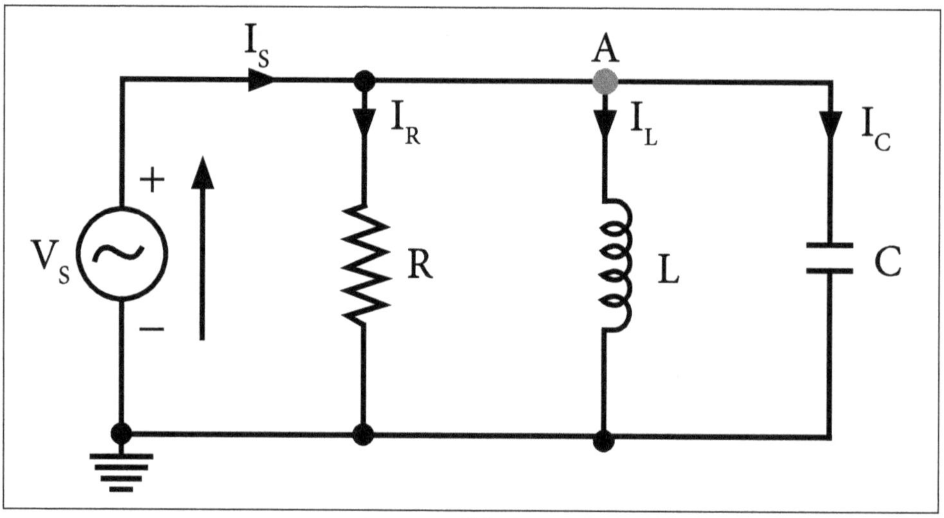

RLC parallel circuit.

The total current, I_S drawn from the supply is equal to the vector sum of the resistive, inductive and capacitive current, not the mathematical sum of the three individual branch currents, as the current flowing in resistor, inductor and capacitor are not in same phase with each other.

Apply Kirchhoff's current law, which states that the sum of the currents entering a junction or node is equal to the sum of the current leaving that node. Thus, we get:

$$I_s^2 = I_R^2 + (I_L - I_C)^2$$

Phasor Diagram of Parallel RLC Circuit

Let,

I_R is the current flowing through the resistor.

I_S is the total source current.

I_L is the current flowing through the inductor.

I_C is the current flowing through the capacitor.

θ is the phase angle difference between the supply voltage and current.

V is the supply voltage.

For drawing the phasor diagram of parallel RLC circuit, voltage is taken as reference because voltage across each element remains the same and all the other currents i.e., I_R, I_C, I_L are drawn relative to this voltage vector.

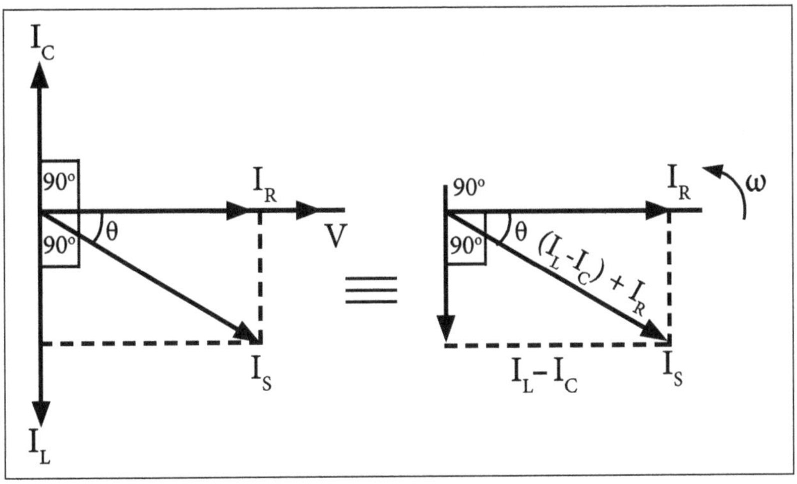

Phasor diagram.

In case of resistor, voltage and current are in same phase. So, draw the current vector I_R in same phase and direction to voltage. For capacitor, current leads the voltage by 90° so, draw I_C vector leading the voltage vector, V by 90°.

For inductor, current vector I_L lags voltage by 90° so draw I_L lagging voltage vector, V by 90°. Now, draw the resultant of I_R, I_C, I_L i.e., current I_S at a phase angle difference of θ with respect to voltage vector, V.

Simplifying the phasor diagram, we get a simplified phasor diagram. On this phasor diagram, we can apply Pythagoras theorem and we get:

$$I_s^2 = I_R^2 + (I_L - I_C)^2$$

Series-Parallel Circuits

Consider the series-parallel circuit consisting of three branches A, B and C as shown in the below figure.

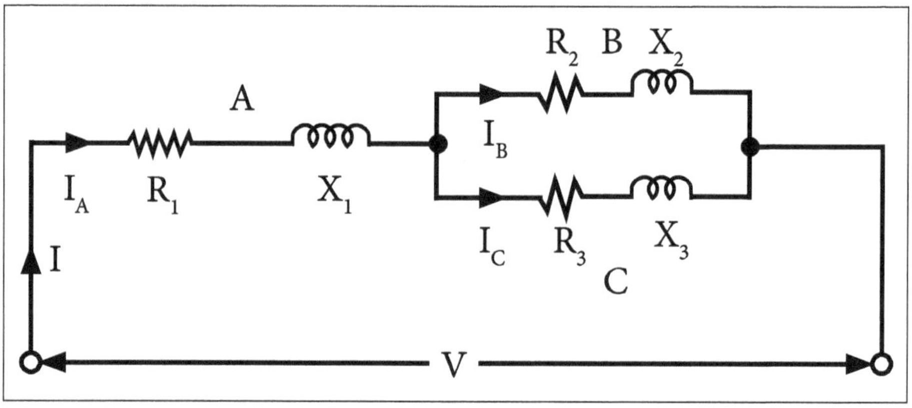

Impedance of branch A, $Z_A = R_1 + jX_1$

Impedance of branch B, $Z_B = R_2 + jX_2$

Impedance of branch C, $Z_C = R_3 + jX_3$

Total impedance of the circuit Z is given by:

$$Z = Z_A + \frac{Z_B Z_C}{Z_B + Z_C}$$

Total current:

$$I = \frac{V}{Z} = I_A$$

Current:

$$I_B = I\frac{Z_C}{Z_B + Z_C}$$ (Applying current divider rule)

$$I_C = I\frac{Z_B}{Z_B + Z_C}$$

By applying admittance method we can also solve the problem as:

$$Y_A = \frac{1}{Z_A}; Y_B = \frac{1}{Z_B}; Y_C = \frac{1}{Z_C}$$

Total admittance of the parallel branches B and C is given by:

$$Y_{BC} = Y_B + Y_C$$

Impedance:

$$Z_{BC} = \frac{1}{Y_{BC}}$$

Total impedance, $Z = Z_A + Z_{BC}$

Total current is given by:

$$I = \frac{V}{Z} = \frac{V}{Z_A + Z_{BC}}$$

3.2.1 Power Factor and Power in AC Circuit

Power Factor

Power Factor is the cosine of the phase angle between current and voltage. i.e.,

$$\text{Power Factor} = \cos\theta$$

In other words, Power Factor is the ratio of true power to apparent power.

$$\text{Power Factor} = \frac{\text{True power}}{\text{Apparent power}}$$

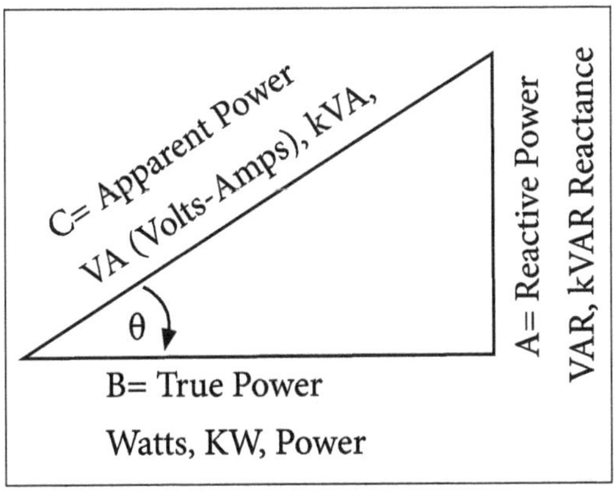

Power triangle.

Power in AC Circuits

50 or 60Hz AC electric power constitutes the most common form of electric power distribution, the phasor notation developed is employed to analyse the power absorbed by both resistive and complex loads.

Instantaneous and Average Power

When a linear electric circuit is excited by a sinusoidal source, all currents and voltages in the circuit are also sinusoids of the same frequency as that of the excitation source. Figure below shows the general form of a linear AC circuit. The most general expressions for the voltage and current delivered to an arbitrary load are,

$$v(t) = V\cos(\omega t - \theta V)$$

$$i(t) = I\cos(\omega t - \theta I) \qquad \text{...(1)}$$

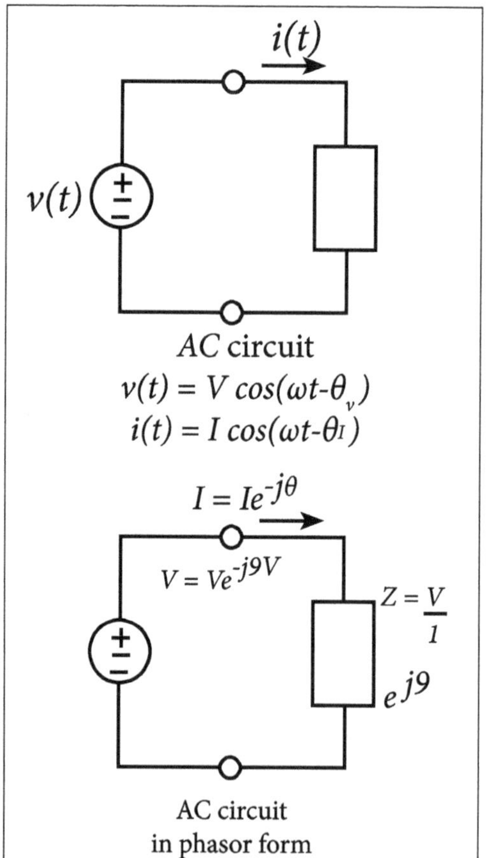

(1) Circuit for illustration of AC power.

Where I and V are the peak amplitudes of the sinusoidal current and voltage, respectively and θ_V and θ_I are their phase angles. Two such waveforms are plotted in the below figure, with unit amplitude and with phase angles $\theta_V = \pi/6$ and $\theta_I = \pi/3$. Therefore, the phase shift between the source and load is $\theta = \theta_V - \theta_I$. Assume that $\theta_V = 0$, without any loss of generality, since all phase angles will be referenced to the source voltage's phase.

Since the instantaneous power dissipated by a circuit element is given by the product of the instantaneous voltage and current, it is possible to obtain a general expression for the power dissipated by an AC circuit element:

$$p(t) = v(t)i(t) = VI\cos(\omega t)\cos(\omega t - \theta) \quad \text{...(2)}$$

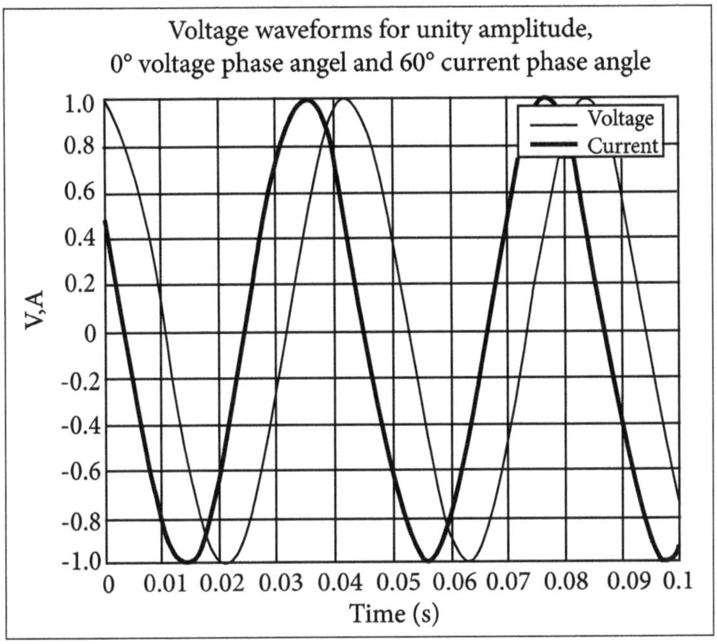

(2) Current and voltage waveforms for illustration of AC power.

Equation (2) can be further simplified with the aid of trigonometric identities to yield:

$$p(t) = \frac{VI}{2}\cos(\theta) + \frac{VI}{2}\cos(2\omega t - \theta) \quad \text{...(3)}$$

Where θ is the difference in phase between voltage and current.

Equation (3) illustrates the instantaneous power dissipated by an AC circuit element is equal to the sum of an average component $1/2\ VI\cos(\theta)$ and a sinusoidal component $1/2\ VI\cos(2\omega t - \theta)$, oscillating at a frequency double that of the original source frequency. The instantaneous and average power are plotted in the below figure for the signals of figure (2). The average power corresponding to the voltage and current signals of equation (1) can be obtained by integrating the instantaneous power over one cycle of the sinusoidal signal. Let $T = 2\pi/\omega$ represent one cycle of the sinusoidal signals. Then the average power P_{av} is given by the integral of the instantaneous power p(t),

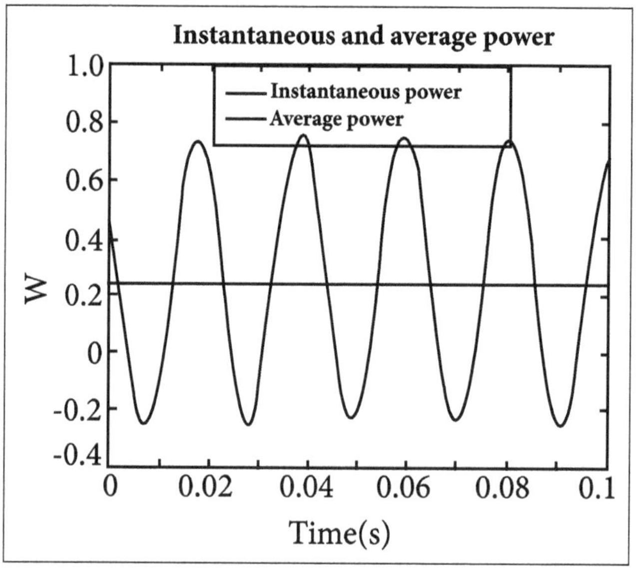

Instantaneous and average power dissipation.

Over one cycle the Average power given by:

$$P_{av} = \frac{1}{T}\int_{o}^{T} p(t)dt \qquad ...(4)$$

$$= \frac{1}{T}\int_{o}^{T} \frac{VI}{2}\cos(\theta)dt + \frac{1}{T}\int_{o}^{T} \frac{VI}{2}\cos(2\omega t - \theta)dt$$

$$P_{av} = \frac{VI}{2}\cos(\theta) \qquad ...(5)$$

Since, the second integral is equal to zero and cos(θ) is a constant.

The same analysis carried out in equations (1) to (3) can also be repeated using phasor analysis. In phasor notation, the current and voltage of equation (1) are given by:

$$V(j\omega) = V\,e^{j\theta} \qquad ...(6)$$

$$I(j\omega) = I\,e^{-j\theta} \qquad ...(7)$$

Note further that the impedance of the circuit element shown in figure (1) is defined by the phasor voltage and current of equations (6) and (7) to be:

$$Z = \frac{V}{I}e^{j(\theta)} = |Z|e^{j\theta} \qquad ...(8)$$

The expression for the average power obtained in equation (4) can therefore also be represented using phasor notation, as follows:

$$P_{av} = \frac{1}{2}\frac{V^2}{|Z|}\cos\theta = \frac{1}{2}I^2|Z|\cos\theta \qquad ...(9)$$

Complex Power

Complex power is the product of voltage and complex conjugate of current, both in phasor form. For inductive circuit, the voltage (V ∠ 0°) is taken as reference and the current (I ∠ -φ = I cos φ - j I sin φ) is lagging the voltage by an angle, φ.

The complex power is given by:

$$\bar{S} = \bar{V}.I = V\angle o^{\circ}.I\angle\phi = (VI)\angle\phi = VI\cos\phi + jVI\sin\phi = P + jQ$$

Consider a single phase network in which the voltage and current can be represented in complex form as V.ejα and I.ejβ. Where α and β are angles that voltage vector and current vector subtend with respect to some reference axis respectively. The active and reactive power can be calculated by finding the product of voltage to the conjugate of current. That means:

$$= \left(Ve^{j\alpha}\right)\times\left(Ie^{j\beta}\right)^*$$

$$= \left(Ve^{j\alpha}\right)\times\left(Ie^{-j\beta}\right)$$

$$= VIe^{j(\alpha-\beta)}$$

This (α – β) is nothing but the angle between voltage and current, hence that is phase difference between voltage and current which is normally denoted as φ.

Therefore, the above equation can be re-written as:

$$S = VI\cos\phi + jVI\sin\phi$$

or,

$$S = P + jQ$$

Where,

$$P = VI\cos\varphi$$

$$Q = VI\sin\varphi.$$

This quantity S is called the complex power. The magnitude of complex power i.e. $|S| = (P^2 + Q^2)^{1/2}$ is known as the apparent power and its unit is volt-ampere. This quantity is a product of absolute value of voltage and current. Again absolute value of current is directly related to heating effect as per Joule's law of heating. Hence, rating of an electrical machine is normally determined by its apparent power carrying capability within allowable temperature limit.

It is noted that in the equation of complex power, the term Q [= VI sinφ] is positive when φ [= (α – β)] is positive that is, current lags the voltage which means the load is inductive in nature. Again Q is negative when φ is negative, that is current leads the voltage that means the load is capacitive.

Three-Phase Power

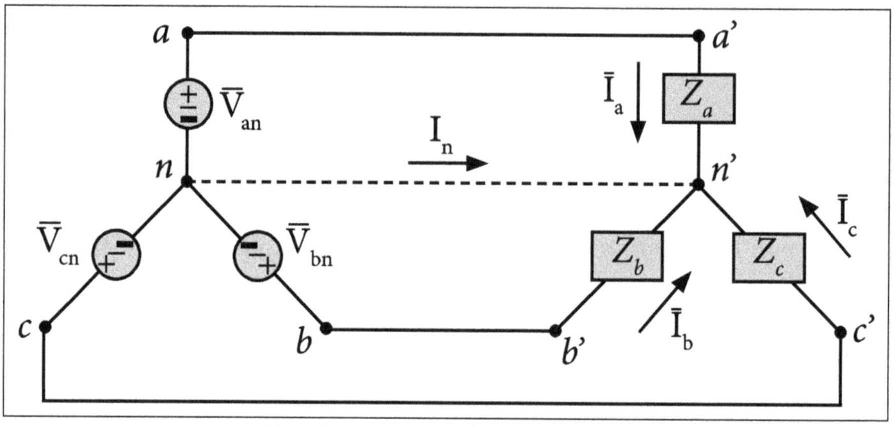

(1) Balanced three-phase AC circuit.

Most of the AC power used today is generated and distributed as three-phase power, by means of an arrangement in which three sinusoidal voltages are generated out of phase with one another. The primary reason is efficiency. The weight of the conductors and other components in a three-phase system is much lower than that in a single-phase system delivering the same amount of power.

Further, while the power produced by a single-phase system has a pulsating nature, a three-phase system can deliver a steady, constant supply of power. For example, a three-phase generator producing three balanced voltages, that is, voltages of equal amplitude and frequency displaced in phase by 120° has the property of delivering constant instantaneous power.

Another important advantage of three-phase power is that, three-phase motors have a nonzero starting torque, unlike their single phase counterpart. The change to three-phase AC power systems from the early DC system proposed by Edison was therefore due to a number of reasons: the efficiency resulting from transforming voltages up and down to minimize transmission losses over long distances, the ability to deliver constant power (an ability not shared by single- and two-phase AC systems), a more efficient use of conductors and the ability to provide starting torque for industrial motors.

Consider a three-phase source connected in the wye (or Y) configuration, as shown in the below figure. Each of the three voltages is 1200 out of phase with the others, so that, using phasor notation, we may write:

$$\bar{V}_{an} = \bar{V}_{an}\angle 0^\circ$$

$$\bar{V}_{bn} = \bar{V}_{bn}\angle -(120^\circ)$$

$$\bar{V}_{cn} = \bar{V}_{cn}\angle(-240^\circ) = \bar{V}_{cn}\angle 120^\circ \qquad ...(1)$$

Where the quantities $\bar{V}_{an}, \bar{V}_{bn}, \text{and}\, \bar{V}_{cn}$ are rms values and are equal to each other. To simplify the notation, it will be assumed:

$$\bar{V}_{an} = \bar{V}_{bn} = \bar{V}_{cn} = \bar{V} \qquad ...(2)$$

In the circuit of figure, the resistive loads are also wye-connected and balanced (i.e., equal). The three AC sources are all connected together at a node called the neutral node, denoted by n. The voltages $\bar{V}_{an}, \bar{V}_{bn}$ and $\bar{V}_{cn}$ are called the phase voltages and form a balanced set in the sense that:

$$\bar{V}_{an} + \bar{V}_{bn} + \bar{V}_{cn} = 0 \qquad ...(3)$$

The sequence of phasor voltages shown in the below figure is usually referred to as the positive (or abc) sequence.

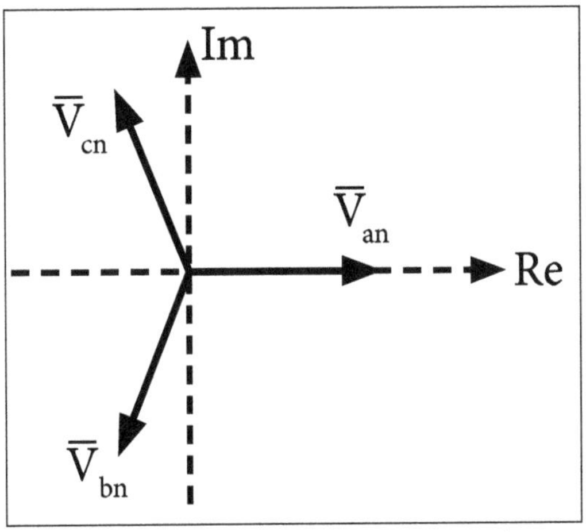

(2) Positive or abc, sequence for balanced three-phase voltages.

Now, consider the "lines" connecting each source to the load and observe that it is possible to also define line voltages (also called line-to-line voltages) by considering the voltages between lines aa' and bb', lines aa' and cc' and lines bb' and cc'. Since the line voltage, say, between aa' and bb' is given by,

$$\bar{V}_{ab} = \bar{V}_{an} + \bar{V}_{nb} = \bar{V}_{an} - \bar{V}_{bn} \qquad ...(4)$$

The line voltages may be computed relative to the phase voltages as,

$$\bar{V}_{ab} = \bar{V}\angle 0° - \bar{V}\angle(-120°) = \sqrt{3}\,\bar{V}\angle 30°$$

$$\bar{V}_{bc} = \bar{V}\angle(-120°) - \bar{V}\angle 120° = \sqrt{3}\bar{V}\angle(-90°)$$

$$\bar{V}_{ca} = \bar{V}\angle 120° - \bar{V}\angle 0° = \sqrt{3}\bar{V}\angle 150° \qquad ...(5)$$

It can be seen that the magnitude of the line voltages is equal to √3 times the magnitude of the phase voltages.

One of the important features of a balanced three phase system is that it does not require a fourth wire, since the current is identically zero (for balanced load $Z_a = Z_b = Z_c = Z$). This can be shown by applying KCL at the neutral node n:

$$\bar{I}_n = \bar{I}_a + \bar{I}_b + \bar{I}_c$$

$$= \frac{1}{Z}\left(\bar{V}_{an} + \bar{V}_{bn} + \bar{V}_{cn}\right)$$

$$= 0 \qquad ...(6)$$

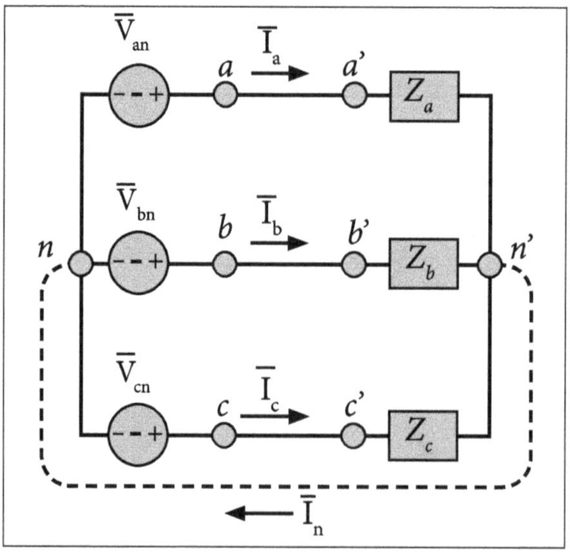

(3) Balanced three-phase AC circuit (redrawn).

Another, more important characteristic of a balanced three-phase power system may be illustrated by simplifying the circuits of figures (1) and (2) by replacing the balanced load impedances with three equal resistances R. With this simplified configuration, one can show that the total power delivered to the balanced load by the three-phase generator is constant. This is an extremely important result: Delivering power in a smooth fashion reduces the wear and stress on the generating equipment.

To show that the total power delivered by the three sources to a balanced resistive load is constant, consider the instantaneous power delivered by each source:

$$p_a(t)=\frac{\bar{V}^2}{R}(1+\cos 2\omega t)$$

$$p_b(t)=\frac{\bar{V}^2}{R}\left[1+\cos(2\omega t-120^\circ)\right]$$

$$p_c(t)=\frac{\bar{V}^2}{R}\left[1+\cos(2\omega t+120^\circ)\right] \quad ...(7)$$

The total instantaneous load power is then given by the sum of the three contributions:

$$p(t)=p_a(t)+p_b(t)+p_c(t)$$

$$=\frac{3\bar{V}^2}{R}+\frac{\bar{V}^2}{R}\left[\cos 2\omega t+\cos(2\omega t-120^\circ)+\cos(2\omega t+120^\circ)\right]$$

$$=\frac{3\bar{V}^2}{R}=\text{constant!} \quad ...(8)$$

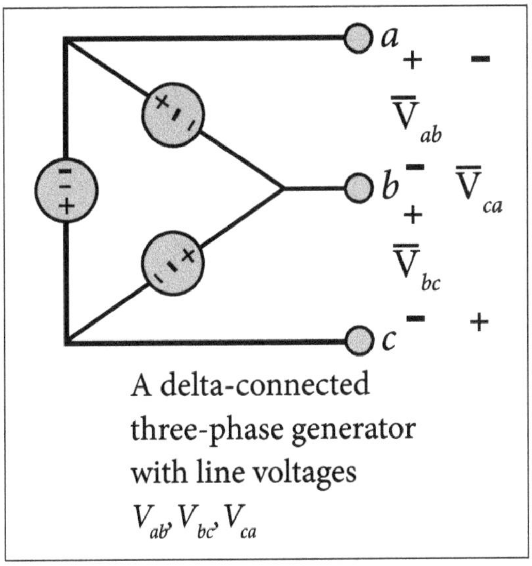

(4) Delta connected generators.

It is also possible to connect the three AC sources in a three-phase system in a delta (or Δ) connection, although in practice this configuration is rarely used.

True Power or Real Power

The mathematical symbol for reactive power is the capital letter Q. The actual amount of power being used or dissipated in a circuit is called true power and it is measured in watts.

Reactive Power

We know that the reactive loads such as capacitors and inductors dissipate zero power, yet the fact that they drop voltage and draw current gives the deceptive impression that they actually do dissipate power. This “phantom power” is called reactive power and it is measured in a unit called Volt-Amps-Reactive (VAR).

Apparent Power

The combination of reactive power and true power is called apparent power and it is the product of a circuit’s voltage and current, without reference to phase angle. Apparent power is measured in Volt-Amps (VA) and is symbolized by the capital letter S.

As a rule, true power is a function of the circuit’s dissipative element, usually resistances (R). Reactive power is a function of a circuit’s reactance (X). Apparent power is a function of a circuit’s total impedance (Z). There are several power equations relating the three types of power to resistance, reactance and impedance (all using scalar quantities). They are as follows:

$P = \text{True power}$

$P = I^2R$

$P = E^2/R$

Measured in units of watts.

$Q = \text{Reactive power}$

$Q = I^2X$

$Q = E^2/X$

Measured in units of volt-amps-reactive (VAR).

$S = \text{Apparent power}$

$S = I^2Z$

$S = E^2/Z$

$S = IE$

Measured in units of volt-amps (VA).

These three types of power such as true, reactive and apparent power relate to one another in trigonometric form. We call this as the power triangle.

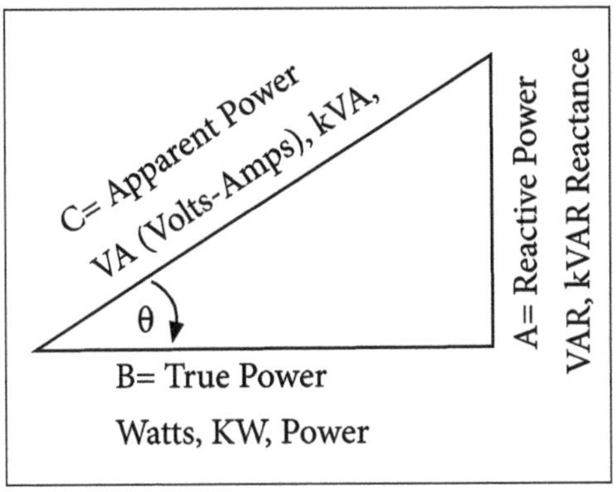

Power triangle.

Problems

1. Let us determine the true power, reactive power and apparent power for the given figure. Assume circuit is purely resistive load.

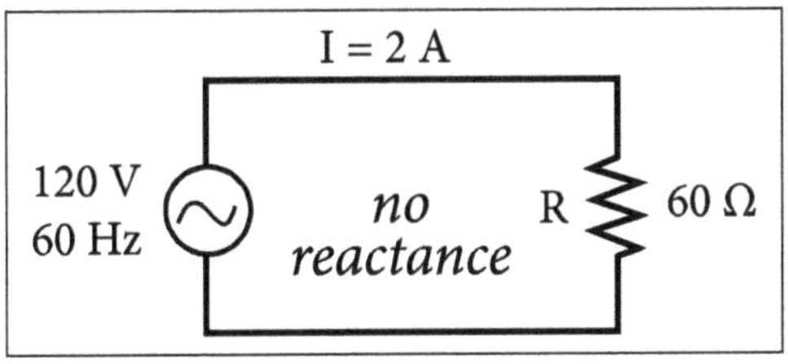

Solution:

Given:

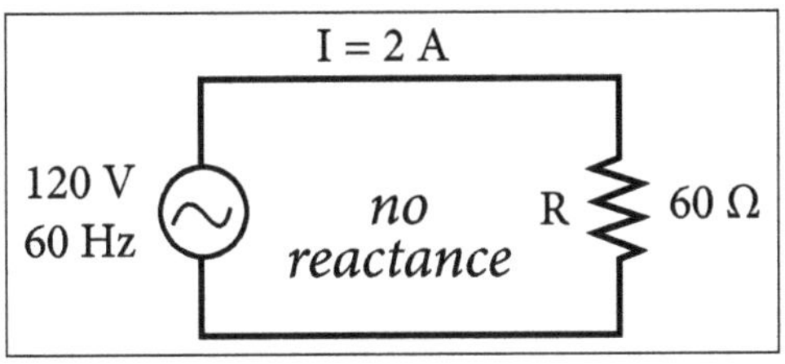

$I = 2\text{ A}$

$V = 120\text{V}$

$f = 60\text{Hz}$

$R = 60\Omega$

$Z = R = 60\Omega$ ($Z = R$ for pure resistive load)

Formula to be used:

$$P = I^2 R$$

$$Q = I^2 X$$

$$S = I^2 Z$$

P = True power = $I^2R = 2^2 \times 60 = 240$ W

Q = Reactive power = $I^2X = 2^2 \times 0 = 0$ VAR

S = Apparent power = $I^2Z = 2^2 \times 60 = 240$ VA

2. Let us determine the true power, reactive power and apparent power for the given figure. Assume circuit is purely reactive load.

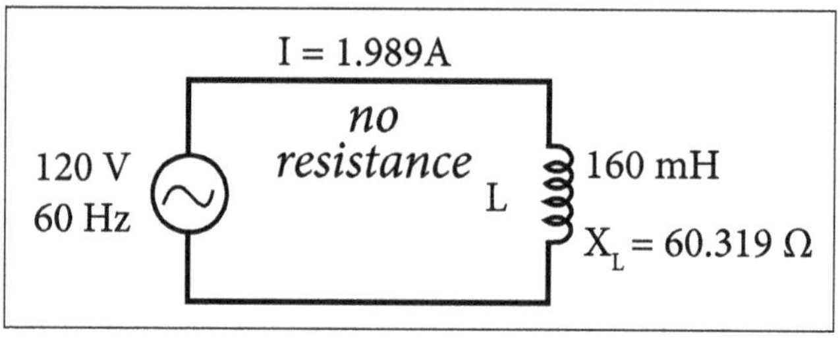

Solution:

Given:

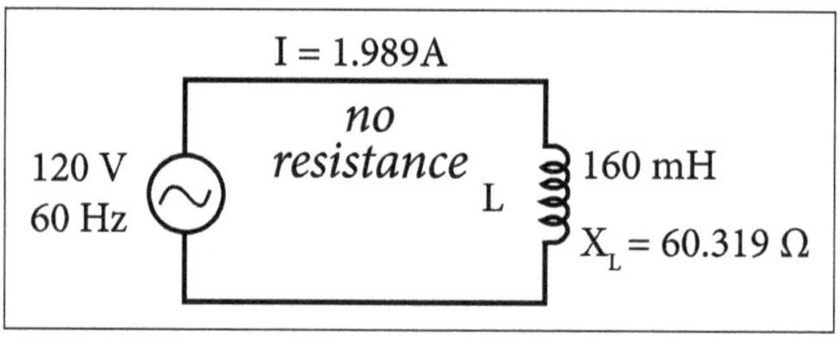

$$I = 1.989 \text{ A}$$

$$V = 120\text{V}$$

$$f = 60 \text{ Hz}$$

$$L = 160\text{mH}$$

$$X_L = 60.319\Omega$$

$Z = X_L = 60.319\Omega$ ($Z = X_L$ for pure reactance load)

Formula to be used:

$$P = I^2R$$

$$Q = I + X$$

$$S = I^2Z$$

P = True power = $I^2R = (1.989)^2 \times 0 = 0$ W

Q = Reactive power = $I^2X = (1.989)^2 \times 60.319 = 238.73$ VAR

S = Apparent power = $I^2Z = (1.989)^2 \times 60.319 = 238.73$ VA

3. Let us find the true power, reactive power and apparent power for the given figure. Assume circuit is purely Resistive/reactive load.

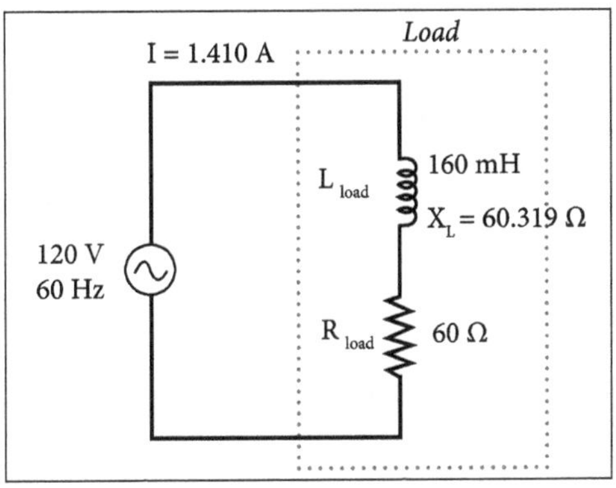

Solution:

Given:

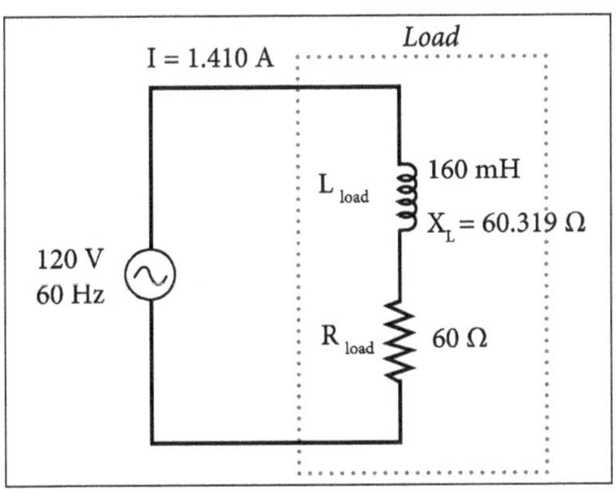

$$I = 1.410 \text{ A}$$

$$V = 120\text{V}$$

$R_{load} = 60\Omega$

$f = 60\ Hz$

$L_{load} = 160\ mH$

$X_L = 60.319\Omega$

$Z = X_L = 60.319\Omega$ ($Z = X_L$ for pure reactance load)

Formula to be used:

$P = I^2R$

$Q = I^2X$

$S = I^2Z$

P = True power = $I\ R = (1.989)\ \times 60 = 119.365\ W$

Q = Reactive power = $I^2X = (1.989)^2 \times 60.319 = 119.998\ VAR$

S = Apparent power = $I^2Z = 169.256\ VA$

3.2.2 Parallel Circuits, Effects of Frequency Variation and Resonance in RLC Series

Resonance Condition

When $X_L = X_C$ [or] the current in the circuit is in-phase with the applied voltage, it is called Resonance condition. In problems, equate imaginary part of the Impedance (Z) or Admittance (Y) to zero to get resonance condition. [i.e., if $Z = Re + jI_m$, then equate $I_m = 0$].

Two types of Resonance Circuits

1. Series Resonance

The basic series-resonant circuit is shown in the below figure. As the frequency of the source changes, the maximum amplitude of the source voltage (V_m) is held constant.

From the figure,

$V_S = V_m \cos(\omega t)$

$i = I_m \cos(\omega t + \theta)$

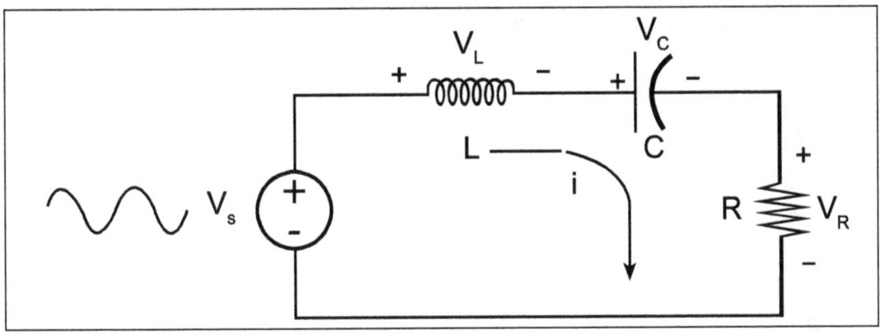

Series resonant circuit.

The frequency at which the reactance of the inductance and the capacitance cancel each other is the resonant frequency of this circuit. This occurs at:

$$\omega_o = 1/\sqrt{(LC)}$$

Since $i = V_R/R$, then the current i can be studied by studying the voltage across the resistor.

The current i has the expression:

$$i = I_m \cos(\omega t + \theta)$$

$$I_m = \frac{V_m}{\sqrt{R^2 + \left(\omega L - \frac{1}{\omega C}\right)^2}}$$

$$\theta = -\tan^{-1}\left(\frac{\omega L - \frac{1}{\omega C}}{R}\right)$$

This yields the bandwidth:

$$B = \omega_2 - \omega_1 = R/L.$$

$$\omega_{2,1} = \sqrt{\left(\frac{R}{L}\right)^2 + \frac{1}{LC}} \pm \frac{R}{2L}$$

$\omega_{2,1}$ is the half power frequency or the 3 dB frequency, (i.e.) the frequencies at which the value of I_m equals the maximum possible value divided by 1.414.

The quality factor:

$$Q = \frac{\omega_o}{B} = \frac{1}{R}\sqrt{\frac{L}{C}}$$

Then the maximum value of V_L occurs at:

$$\omega = \frac{\omega_0}{\sqrt{1 - \frac{R^2C}{2L}}}$$

V_C occurs at:

$$\omega_0\sqrt{1 - \frac{R^2C}{2L}}$$

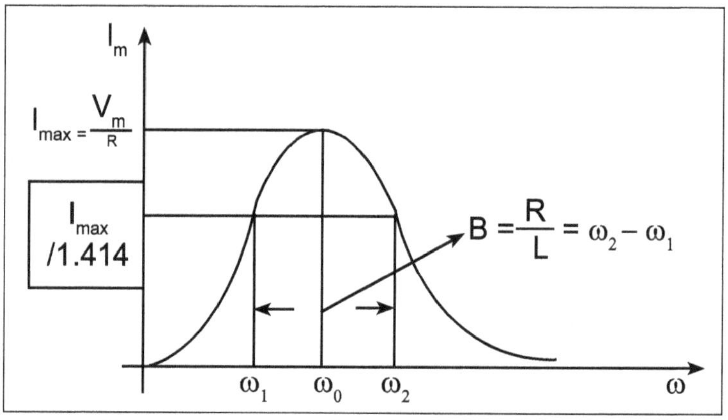

Frequency response of a Series - Resonant Circuit.

2. Parallel Resonance

The basic parallel-resonant circuit is shown in the below figure. The steady state amplitude and the phase angle of the output voltage V_o vary with the frequency of the sinusoidal voltage source.

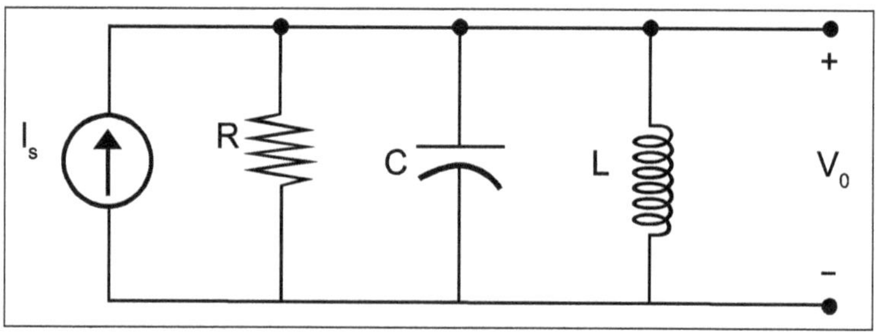

Parallel resonant circuit.

From the figure,

$$I_s = I_m \cos(\omega t)$$

$$V_o = V_m \cos(\omega t + \theta)$$

If $I_s = I_m\cos(\omega t)$, then $V_o = V_m\cos(\omega t+\theta)$

Where,

$$V_m = \frac{I_m}{\sqrt{\frac{1}{R^2}+\left(\omega C-\frac{1}{\omega L}\right)^2}}$$

$$\text{And } \theta = -\tan^{-1}\left(R\left(\omega C-\frac{1}{\omega L}\right)\right)$$

The resonant frequency is:

$$\omega_o = \frac{1}{\sqrt{LC}}$$

The 3 dB frequencies are:

$$\omega_{2,1} = \sqrt{\left(\frac{1}{2RC}\right)^2+\frac{1}{LC}} \pm \frac{1}{2RC}$$

The bandwidth:

$$B = \omega_2 - \omega_1 = 1/RC$$

The quality factor:

$$Q = \omega_o / B = R\sqrt{(C/L)}$$

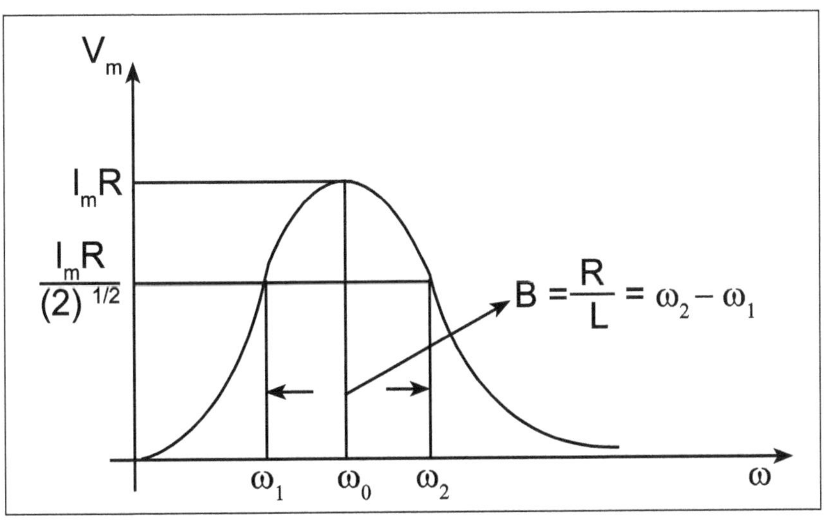

Frequency Response of Parallel Resonant Circuit.

Terms affected by frequency	R Independent of frequency		$jX_L = j\omega_L$ Directly proportional to frequency		$-jX_c = -j/\omega_c$ Inversely proportional to frequency	
Frequency	Magnitude $\|M\| = R$	Phase	Magnitude $\|M\| = \omega_L$	Phase = $\tan^{-1}(I_m/R)$	Magnitude $\|M\| = 1/\omega C$	Phase = $\tan^{-1}(I_m/R)$
$\omega = 0$	R	0°	$\|M\| = 0$	$\phi = \tan^{-1}(\omega_L/0)$ $\phi = \tan^{-1}(\infty)$ $\phi = 90^\circ$	$\|M\| = -\infty$	$\phi = -\tan^{-1}(\omega_C/0)$ $\phi = -\tan^{-1}(\infty)$ $\phi = -90^\circ$
$\omega = \infty$	R	0°	$\|M\| = \infty$		$\|M\| = -0$	

Series RLC Circuit - Resonance Frequency (fr)

In a series RLC circuit, series resonance occurs when $X_L = X_C$. The frequency at which the resonance occurs is called resonant frequency.

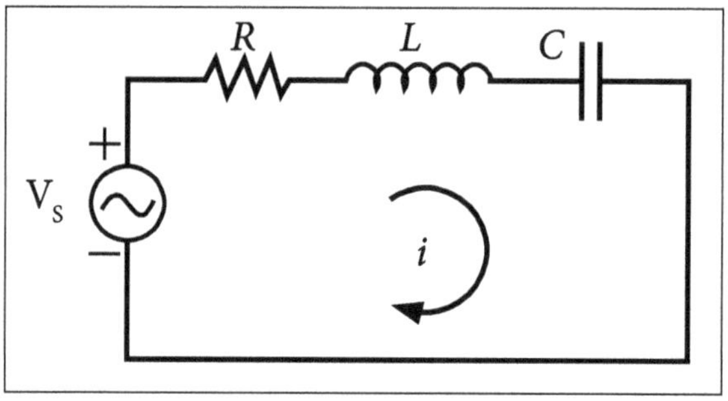

At resonance:

$$X_L = X_C$$

$$\omega_r L = \frac{1}{\omega_r C} \Rightarrow 2\pi f_r L = \frac{1}{2\pi f_r C}$$

$$4\pi^2 f_r^2 LC = 1$$

$$f_r^2 = \frac{1}{4\pi^2 LC}$$

Resonance frequency:

$$f_r = \frac{1}{2\pi\sqrt{LC}} \text{Hz}$$

[or]

$$\omega_r = \frac{1}{\sqrt{LC}} \text{rad/sec}$$

Total impedance of the RLC circuit:

$$Z = R + jX_L - jX_c$$

$$Z = R - j\left(\frac{1}{\omega C} - \omega L\right)$$

$$\text{Magnitude}|Z| = \sqrt{R^2 + \left(\frac{1}{\omega C} - \omega L\right)^2}$$

$$\text{Phase}\,\angle\varphi = \tan^{-1}\left(\frac{\frac{1}{\omega C} - \omega L}{R}\right) = -\tan^{-1}\left(\frac{1}{\omega CR} - \frac{\omega L}{R}\right)$$

At resonance, Magnitude $|Z| = \sqrt{R^2} = R$

Current flowing through RLC circuit, $I = V/|Z|$

At resonant frequency f_r, $I = V/R$

Voltage across Resistor, $V_R = I\,R$

Voltage across Inductor, $V_L = I\,X_L$ (or) $I{*}\omega_L$

Voltage across Capacitor, $V_C = I\,X_C$ (or) $I/\,\omega_c$

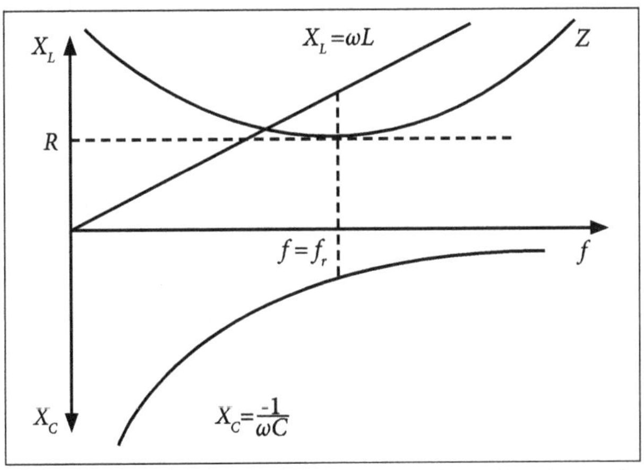

Frequency response of Impedance Z, R, X_L and X_C of series RLC circuit.

Maximum voltage occurs across the inductor above the resonant frequency:

$$f_L = \frac{1}{2\pi\sqrt{LC}}\sqrt{\frac{1}{1 - \frac{R^2C}{2L}}}$$

Maximum voltage occurs across the capacitor below the resonant frequency:

$$f_C = \frac{1}{2\pi}\sqrt{\frac{1}{LC} - \frac{R^2}{2L}}$$

Frequency Response of Resistor, Inductor and Capacitor - Series RLC circuit:

$$\text{Magnitude}\,|Z| = \sqrt{R^2 + \left(\frac{1}{\omega C} - \omega L\right)^2}$$

From the above figure:

- The reactance of capacitance X_C varies inversely with frequency.
- The reactance of X_L varies directly proportional to frequency.
- The resistor is independent of frequency variation.

Impedance and current vs. frequency in series resonant circuit:

At resonant frequency, the capacitive reactance is equal to the inductive reactance and hence, the impedance is minimum. Because of minimum impedance, maximum current flows through the circuit.

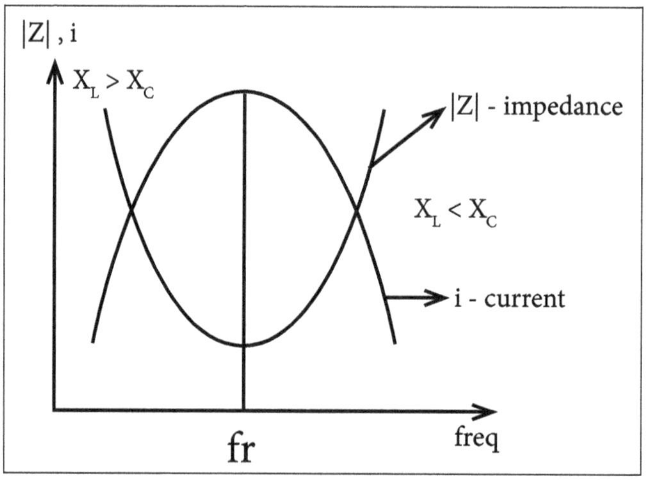

Frequency response of current & impedance.

From the figure:

Case (i): $X_L > X_C$: As frequency increases, X_L increases, X_C decreases which causes, increase in current and decrease in impedance.

Case (ii): $X_L = X_C$: As frequency f = resonant frequency f_r, current reaches its maximum and impedance reaches its minimum.

Case (iii): $X_L < X_C$: As frequency increases, X_L increases, X_C also increases which causes, increase in impedance and decrease in current.

Parallel Resonant - RLC Circuit

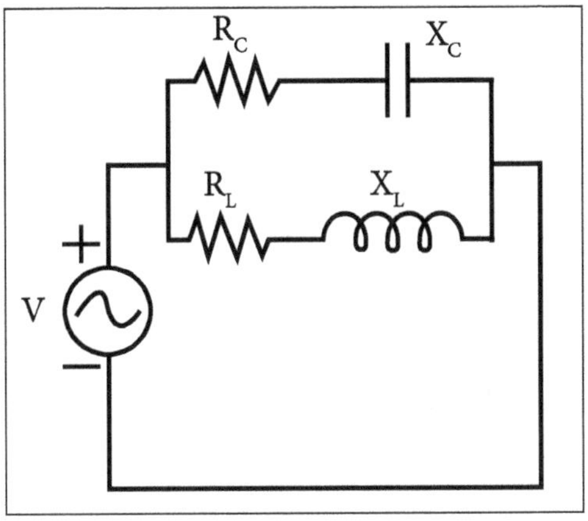

Equate imaginary part equal to zero or at resonance, susceptance part =o [Y = G + jB]

$$\frac{-\omega_r L}{R_L^2 + (\omega_r L)^2} + \frac{\frac{1}{\omega_r C}}{R_C^2 + \left(\frac{1}{\omega_r C}\right)^2}$$

$$\Rightarrow \frac{\frac{1}{\omega_r C}}{R_C^2 + \left(\frac{1}{\omega_r C}\right)^2} = \frac{\omega_r L}{R_L^2 + (\omega_r L)^2}$$

$$= \omega_r L\left[R_C^2 + \left(\frac{1}{\omega_r C}\right)^2\right] = \frac{1}{\omega_r C}\left[R_L^2 + (\omega_r L)^2\right]$$

$$= \omega_r^2\left[R_C^2 + \left(\frac{1}{\omega_r C}\right)^2\right] = \frac{1}{LC}\left[R_L^2 + (\omega_r L)^2\right]$$

$$= \omega_r^2 R_C^2 + \frac{\cancel{\omega}_r^2}{\cancel{\omega}_r^2 R^2} = \frac{1}{LC} R_L^2 + \frac{\omega_r^2 \cancel{L}^2}{\cancel{L} C}$$

$$= \omega_r^2 R_C^2 + \frac{1}{C^2} = \frac{1}{LC} R_L^2 + \frac{\omega_r^2 L}{C}$$

It occurs when $X_L = X_C$. The frequency at which resonance occurs is called the resonant frequency. At resonance, the two branch currents are equal in magnitude and 180° out of phase with each other.

From the circuit, the total admittance is:

$$Y = \frac{1}{R_L + jX_L} + \frac{1}{R_C + jX_C} = \frac{1}{R_L + j\omega L} + \frac{1}{R_C - \frac{j}{\omega C}}$$

Multiply both Nr & Dr by complex conjugate:

$$Y = \frac{1}{R_L + j\omega L} \times \frac{R_L - j\omega L}{R_L - j\omega L} + \frac{1}{R_C - \frac{j}{\omega C}} \times \frac{R_C + \frac{j}{\omega C}}{R_C + \frac{j}{\omega C}}$$

$$Y = \frac{R_L - j\omega L}{R_L^2 + (\omega L)^2} + \frac{R_C + \frac{j}{\omega C}}{R_C^2 + \left(\frac{1}{\omega C}\right)^2}$$

$$Y = \frac{R_L}{R_L^2 + (\omega L)^2} + \frac{R_C}{R_C^2 + \left(\frac{1}{\omega C}\right)^2} + j\left[\frac{-\omega L}{R_L^2 + (\omega L)^2} + \frac{\frac{1}{\omega C}}{R_C^2 + \left(\frac{1}{\omega C}\right)^2}\right]$$

$$\omega_r^2 R_C^2 - \frac{\omega_r^2 L}{C} = \frac{1}{LC} R_L^2 - \frac{1}{C^2}$$

$$\omega_r^2\left[R_C^2 - \frac{L}{C}\right] = \frac{1}{LC}\left[R_L^2 - \frac{L}{C}\right]$$

$$\omega_r = \frac{1}{\sqrt{LC}}\sqrt{\frac{R_L^2 - (L/C)}{R_C^2 - (L/C)}}$$

The condition for resonance frequency is given by substituting $R_L = R_C$:

$$\omega_r = \frac{1}{\sqrt{LC}}$$

$$f_r = \frac{1}{2\pi\sqrt{LC}}.$$

3.3 Q-Factor

Quality Factor, Q

Reactive components such as capacitors and inductors are often described with a figure of merit called Q. While it can be defined in many ways, it's most fundamental description is:

$$Q = \omega \frac{\text{energy stored}}{\text{average power dissipated}}$$

In other words, the quality factor Q, is the ratio of the reactive power in the inductor or capacitor to the true power in the resistance in series with the coil or capacitor:

$$Q = 2\pi \times \frac{\text{max energy stored}}{\text{energy dissipated per cycle}}$$

$$\text{Quality factor of the capacitance circuit, } Q = \frac{1}{wCR}$$

Thus, it is a measure of the ratio of stored vs. lost energy per unit time. Note that this definition does not specify what type of system is required. Thus, it is quite general.

An ideal reactive component (capacitor or inductor) stores energy:

$$E = \frac{1}{2} CV_{pk}^2 \text{ or } \frac{1}{2} LI_{pk}^2$$

Since any real component also has loss due to the resistive component, the average power dissipated is:

$$P_{avg} = \frac{1}{2} I_{ok}^2 R = \frac{V_{pk}^2}{2R}$$

Here we consider an example of a series resonant circuit.

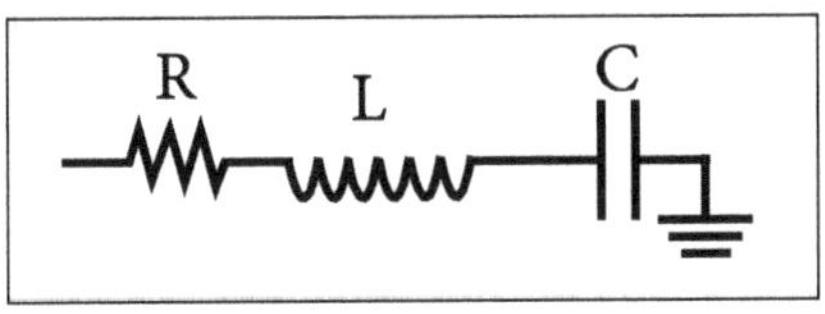

Series - Resonant circuit.

At resonance, the reactance cancels out leaving just a peak voltage, V_{pk} across the loss resistance, R. Thus, $I_{pk} = V_{pk}/R$ is the maximum current which passes through all elements. Then:

$$Q = \omega_0 \frac{LI_{pk}^2 / 2}{I_{pk}^2 R / 2} = \frac{\omega_0 L}{R} = \frac{1}{\omega_0 RC}$$

$$Q = \frac{1}{\omega RC} = \frac{X}{R}$$

Where we pretend that the capacitor is resonated with an ideal inductor at frequency ω.

X is the capacitive reactance and R is the series resistance. Since this Q refers only to the capacitor itself, in isolation from the rest of the circuit, it is called unloaded Q or Q_U. The higher the unloaded Q, the lower the loss. Notice that the Q decreases with frequency.

The unloaded Q of an inductor is given by:

$$Q_U = \frac{\omega_0 L}{R}$$

Where, R is a series resistance.

Note that Q is proportional to frequency for an inductor. The Q of an inductor will depend upon the wire diameter, core material (air, powdered iron, ferrite) and whether it is in a shielded metal can or not. It is easy to show that for a parallel resonant circuit, Q is given by susceptance/conductance:

$$Q = B/G$$

Where,

B is the susceptance of the capacitor or inductor.

G is the shunt conductance.

Quality factor for series and parallel RLC circuit:

Circuit	Bandwidth	Capacitive	Inductive
Series RLC circuit, Quality factor	$Q = \frac{f_r}{f_2 - f_1} = \frac{f_r}{BW} = \frac{f_r}{R/2\pi L}$	$Q = \frac{1}{\omega CR}$	$Q = \frac{\omega L}{R}$
Parallel RLC circuit, Quality factor	$Q = \frac{f_r}{f_2 - f_1} = \frac{f_r}{BW} = \frac{f_r}{1/RC} = f_r RC$	$Q = \omega CR$	$Q = \frac{R}{\omega L}$

3.3.1 Bandwidth of Resonant Circuit

Bandwidth is defined as a band containing all frequencies between upper cut-off and lower cut-off frequencies.

At resonance, the voltages across the capacitor and the pure inductance cancel out, so the series impedance takes its minimum value, $Z_0 = R$. Thus, if we keep the voltage constant, the current is a maximum at resonance. The current goes to zero at low frequency, because X_C becomes infinite. The current also goes to zero at high frequency because X_L increases with ω.

The bandwidth B is defined as the difference between the two frequencies ω_1 and ω_2 at which the circuit converts power at half the maximum rate.

Bandwidth, B = $\omega_2 - \omega_1$

At a certain frequency, the power dissipated by the resistor is half of the maximum power which as mentioned occurs at:

$$\omega_o = 1/\sqrt{(LC)}$$

The half power occurs at the frequencies for which the amplitude of the voltage across the resistor becomes equal to $1/\sqrt{2}$ of the maximum.

$$P_{1/2} = \frac{1}{4}\frac{V_{max}^2}{R}$$

Figure below shows the graphical form of the various frequencies of interest.

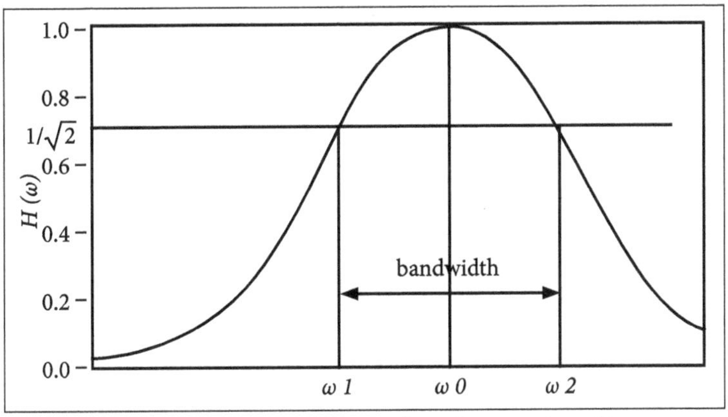

Graphical representation of the various frequencies of interest.

Therefore, the ½ power occurs at the frequencies for which:

$$\frac{1}{\sqrt{2}} = \frac{\omega RC}{(1-\omega^2 LC)^2 + (\omega RC)^2}$$

Equation has two roots:

$$\omega_1 = -\frac{R}{2L} + \sqrt{\left(\frac{R}{2L}\right)^2 + \frac{1}{\omega_o^2}}$$

$$\omega_2 = \frac{R}{2L} + \sqrt{\left(\frac{R}{2L}\right)^2 + \frac{1}{\omega_o^2}}.$$

Bandwidth, $B = \omega_2 - \omega_1$

By multiplying these two equations, we can show that ω_o is the geometric mean of ω_1 and ω_2.

$$\omega_o = \sqrt{(\omega_1\omega_2)}$$

As we see from the plot on the graph, the bandwidth increases with increasing R. Equivalently, the sharpness of the resonance increases with decreasing R.

For a fixed L and C, decrease in R corresponds to a narrower resonance and thus, a higher selectivity regarding the frequency range can be passed by the circuit.

As we increase R, the frequency range over which the dissipative characteristics dominate the behaviour of the circuit increases. In order to quantify this behaviour, we define a parameter called Quality Factor Q which is related to the sharpness of the peak and it is given by:

$$Q = 2\pi \frac{\text{maximum energy stored}}{\text{total energy lost per cycle at resonance}} = 2\pi \frac{E_S}{E_D}$$

Which represents the ratio of the energy stored to the energy dissipated in a circuit.

The energy stored in the circuit is given by:

$$E_S = \frac{1}{2}LI^2 + \frac{1}{2}CVc^2$$

For $V_c = A \sin\omega t$, the current flowing in the circuit is expressed as:

$$I = C\frac{dVc}{dt} = \omega CA\cos(\omega t)$$

The total energy stored in the reactive element is given by:

$$E_S = \frac{1}{2}L\omega^2C^2A^2\cos^2(\omega t) + \frac{1}{2}CA^2\sin^2(\omega t)$$

At the resonance frequency where:

$$\omega = \omega_o = \frac{1}{\sqrt{LC}}$$

The energy stored in the circuit becomes:

$$E_S = \frac{1}{2}CA^2$$

The energy dissipated per period is equal to the average resistive power dissipated times the oscillation period.

i.e.,

$$E_D = R\langle I^2\rangle\frac{2\pi}{\omega_0} = R\left(\frac{\omega_0^2 C^2 A^2}{2}\right)\frac{2\pi}{\omega_0} = 2\pi\left(\frac{1}{2}\frac{RC}{\omega_0 L}A^2\right)$$

And so, the ratio Q becomes,

$$Q = \omega_0 L / R$$

By combining these equations, we obtain the relationship between the bandwidth and the Q factor.

Selectivity

The ratio of bandwidth to the resonant frequency is called selectivity.

That is:

$$S = \frac{B.\omega}{f_r}$$

Comparison of RLC resonant circuits:

Catego-ries	Series	Parallel
Circuit	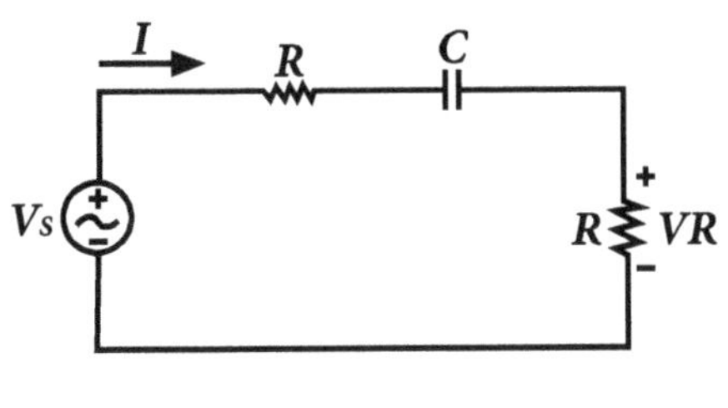	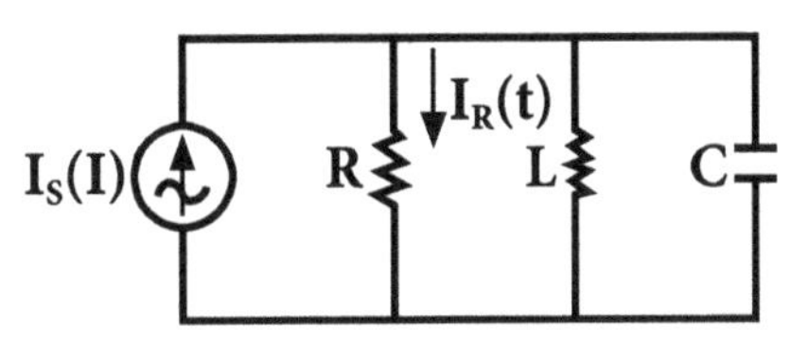
Transfer function	$\lvert H(\omega)\rvert \equiv \left\lvert\frac{V_R}{V_S}\right\rvert = \frac{\omega RC}{\sqrt{(1-\omega^2 LC)^2 + (\omega RC)^2}}$	$\lvert H(\omega)\rvert \equiv \left\lvert\frac{I_R}{I_S}\right\rvert = \frac{\omega L}{\sqrt{(R-\omega^2 LCR)^2 + (\omega L)^2}}$
Resonant frequency	$\omega_0 = \frac{1}{\sqrt{LC}}$	$\omega_0 = \frac{1}{\sqrt{LC}}$

1/2 power frequencies	$\omega_1 = -\frac{R}{2L} + \sqrt{\left(\frac{R}{2L}\right)^2 + \frac{1}{\omega_o^2}}$ $\omega_2 = \frac{R}{2L} + \sqrt{\left(\frac{R}{2L}\right)^2 + \frac{1}{\omega_o^2}}$	$\omega_1 = -\frac{1}{2RC} + \sqrt{\left(\frac{1}{2RC}\right)^2 + \frac{1}{\omega_o^2}}$ $\omega_2 = \frac{1}{2RC} + \sqrt{\left(\frac{1}{2RC}\right)^2 + \frac{1}{\omega_o^2}}$
Bandwidth	$B_S = \omega_2 - \omega_1 = \frac{R}{L}$	$B_r = \omega_2 - \omega_1 = \frac{1}{RC}$
Q factor	$Q = \frac{\omega_o}{B_S} = \frac{\omega_o L}{R} = \frac{1}{\omega_o RC}$	$Q = \frac{\omega_o}{B_r} = \omega_o RC = \frac{R}{\omega_o L}$

4

Semiconductors

4.1 Crystalline Material: Mechanical Properties

A crystal is the solid material in which the component atoms are arranged in a definite pattern and whose surface regularity reflects its internal symmetry. Most of the materials used in electronic devices are crystalline in nature. A three dimensional lattice is defined by the three fundamental translation vectors such that the atomic arrangements look the same in every respect when viewed from two different directions.

Semiconductor Materials

Classification	Example
Gr - IV	Si, Ge
Gr III - V	GaAs, InP, InSb, GaN, GaSb
Gr II - VI	CdS, CdTe, CdSe, ZnO, ZnS
Alloys	$Al_xGa_{1-x}As$, $Ga_xIn_{1-x}As_{1-y}P_y$

A pure crystal is constructed in such a way that it maintains the directional invariance. Thus, the lattice is a regular periodic point in space and is a mathematical abstraction. Such a structure yields a periodic potential throughout the material. An ideal crystal is constructed by the infinite repetition of identical structural unit in space. The crystals are made up of identical building blocks, each block being an atom or a group of atoms.

The periodicity of the crystal is the key property which controls the properties of the carriers inside the materials. The intrinsic property of a crystal is that the environment around the given atom or atoms is identical to that of the other atom or atoms. Mathematically, the crystal structure is the sum of the lattice and the basis, where the basis is the site of the atom attached identically to every lattice point.

Two properties of crystals are of particular interest, since they are the ingredients required to formulate the total current in semiconductors. Firstly, we need to know how many fixed and how many mobile charges are present in the material. Secondly, we should understand the process of transport of the mobile carriers through the semiconductors. The below table illustrates a few elemental and compound semiconductors together with their alloys which find extensive applications in modern semiconductor industries.

It is important to note that the elemental semiconductor Silicon (Si) is preferred over another elemental semiconductor Germanium (Ge) for the following reasons:

- Higher temperature stability.
- Higher band gap.
- Lower leakage current.
- Higher breakdown voltage.
- Technically easier fabrication process than germanium.

The compound III-V semiconductor Gallium Arsenide is preferred over the elemental semiconductor Silicon for the following reasons:

- Higher temperature stability.
- Higher band gap.
- Higher cut-off voltage than silicon.

There is one disadvantage; the fabrication of GaAs is technically more difficult than that of silicon. Some special microelectronic techniques are used to fabricate GaAs based devices.

Mechanical Properties

In the presence of stress, the tensorial nature of Hooke's law becomes prominent. Only the second and third order elastic constants are being affected in the presence of stress. In n-type germanium, the shear elastic constant is being reduced by 8%. In view of large changes of elastic constants with carrier density, the experiments on velocity of sound involving shear mode as a function of carrier concentration exhibit the contribution of electrons to mechanical properties.

4.1.1 Energy Band Theory

In a single isolated atom, the electrons in each orbit have some definite energy associated with it. But in case of solids all the atoms are close to each other, so the energy levels of the outermost orbit electrons are affected by the neighbouring atoms.

When two single or isolated atoms are brought close to each other than the outermost orbit electrons of two atoms are interacted or shared with each other, that is, the electrons in the outermost orbit of one atom experiences a attractive force from the neighbouring or nearest atomic nucleus. Due to this the energies of the electrons will not be in same level, the energy levels of the electrons are changed to a value which is higher or lower than that of the original energy level of the electron.

The electrons in the same orbit exhibit different energy levels. The grouping of these different energy levels is termed as energy band.

However, the energy levels of the inner orbit electrons are not much affected by the presence of neighbouring atoms.

Important Energy Bands in Solids

There are many number of energy bands in solids but three of them are very important. These three energy bands are important to understand the behaviour of solids. These energy bands are:

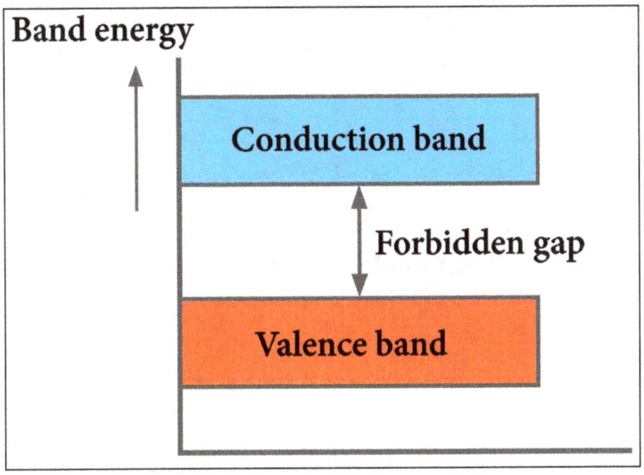

Energy band theory in solids.

Valence band: The energy band which is formed due to grouping the range of energy levels of the valence electrons or outermost orbit electrons is called as valence band. Valence band is present below the conduction band as shown in figure. Electrons in the valence band have lower energy than electrons in conduction band.

The electrons present in the valence band are loosely bound to the nucleus of the atom.

Conduction band: The energy band which is formed due to grouping the range of energy levels of the free electrons is known as conduction band. In general, the conduction band is empty but when external energy is applied the electrons in the valence band jumps to the conduction band and becomes free electrons. Electrons in the conduction band have higher energy than the electrons in valence band. Conduction band electrons are not bound to the nucleus of atom.

4.1.2 Fermi Levels

Fermi Level in Intrinsic Semiconductor

Probability of occupation of energy levels in valence band and conduction band is known as Fermi level. At absolute zero temperature intrinsic semiconductor acts as perfect insulator. However, as the temperature raises free electrons and holes get generated.

In the intrinsic or the pure semiconductor, the number of holes in valence band is equal to the number of electrons in the conduction band. Hence, the probability of occupation of energy levels in conduction band and valence band are equal. Therefore, the Fermi level for the intrinsic semiconductor lies in the middle of forbidden band.

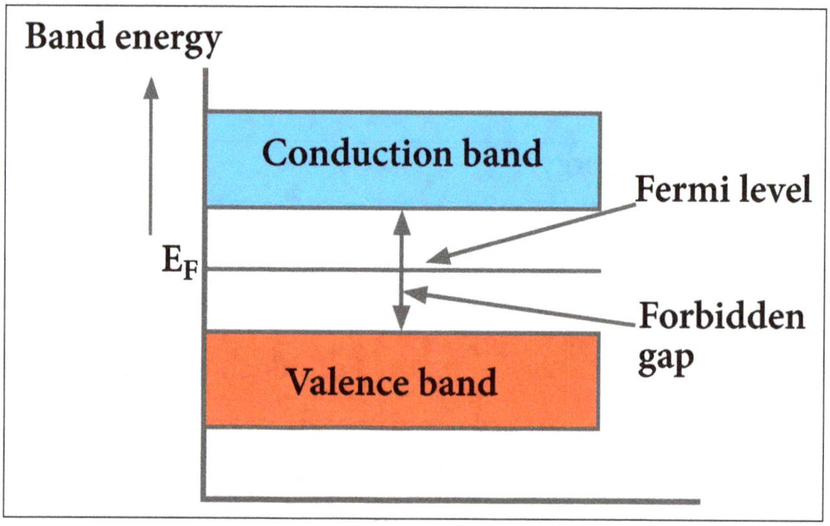

The Fermi level in the middle of forbidden band denotes the equal concentration of free electrons and holes.

The hole-concentration in the valence band is given as:

$$p = N_V e \frac{-(E_F - E_V)}{K_B T}$$

The electron-concentration in the conduction band is given as:

$$n = N_C e \frac{-(E_C - E_F)}{K_B T}$$

Where,

K_B is the Boltzmann constant.

N_c is the effective density of states in the conduction band.

T is the absolute temperature of the intrinsic semiconductor.

N_v is the effective density of states in the valence band.

The number of electrons in the conduction band is depends on the effective density of states in the conduction band and the distance of Fermi level from the conduction band.

The number of holes in the valence band is depends on effective density of states in the valence band and the distance of Fermi level from the valence band.

For an intrinsic semiconductor, the electron-carrier concentration is equal to the hole-carrier concentration.

It can be written as:

$$p = n = n_i$$

Where,

P = hole-carrier concentration.

n = electron-carrier concentration.

n_i = intrinsic carrier concentration.

The fermi level for intrinsic semiconductor is given as:

$$E_F = (E_C + E_V)/2$$

Where,

E_F is the fermi level.

E_C is the conduction band.

E_V is the valence band.

Therefore, the Fermi level in an intrinsic semiconductor lies in the middle of the forbidden gap.

Fermi Level in Extrinsic Semiconductor

In the extrinsic semiconductor, the number of electrons in the conduction band and the number of holes in the valence band are not equal. Hence, the probability of the occupation of energy levels in conduction band and valence band are not equal. Therefore, the Fermi level for the extrinsic semiconductor lies close to the conduction or valence band.

Fermi level in n-type semiconductor: In the n-type semiconductor pentavalent impurity is added. Each pentavalent impurity donates a free electron. The addition of pentavalent impurity produces large number of free electrons in the conduction band.

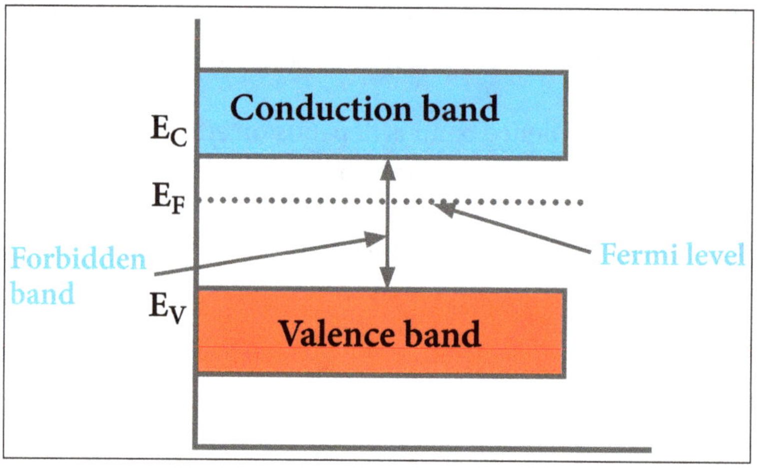

At room temperature, the number of electrons in the conduction band is higher than the number of holes in the valence band. Hence, the probability of occupation of energy levels by the electrons in the conduction band is higher than the probability of occupation of energy levels by the holes in the valence band. This probability of occupation of energy levels is shown in terms of Fermi level. Therefore, the Fermi level in the n-type semiconductor lies close to the conduction band.

The Fermi level for n-type semiconductor is given as:

$$E_F = E_C - K_B T \log \frac{N_C}{N_D}$$

Where,

E_F - The fermi level.

E_C - The conduction band.

K_B - The Boltzmann constant.

T - The absolute temperature.

N_C - The effective density of the states in the conduction band.

N_D - The concentration of donor atoms.

Fermi level in p-type semiconductor: In p-type semiconductor trivalent impurity is added. Each trivalent impurity creates a hole in the valence band and ready to accept an electron. The addition of trivalent impurity creates large number of holes in the valence band.

At room temperature, the number of holes in the valence band is greater than the number of electrons in the conduction band. Hence, the probability of occupation of energy levels by the holes in the valence band is greater than the probability of occupation of energy levels by the electrons in the conduction band. This probability of occupation of energy levels is represented in terms of Fermi level. Therefore, the Fermi level in the p-type semiconductor lies close to the valence band.

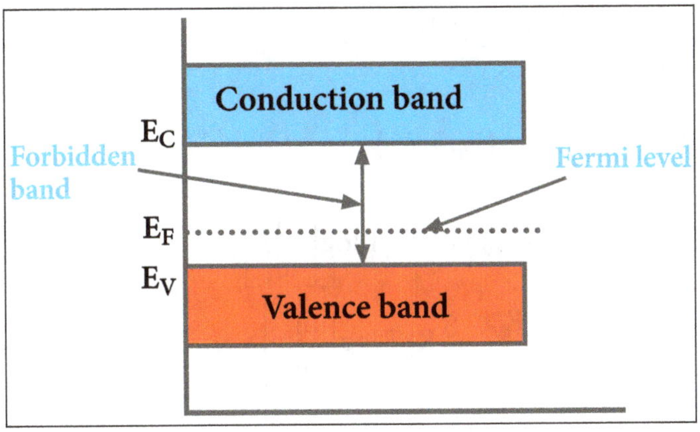

The Fermi level for p-type semiconductor is given as:

$$E_F = E_V + K_B T \log \frac{N_V}{N_A}$$

Where,

N_V is the effective density of states in the valence band.

N_A is the concentration of acceptor atoms.

Derivation of Fermi Level

In an intrinsic semiconductor,

$$2\left[\frac{2\pi m_e^* K_B T}{h^2}\right]^{3/2} \exp\left[\frac{E_F - E_C}{K_B T}\right] = 2\left[\frac{2\pi m_h^* K_B T}{h^2}\right]^{3/2} \exp\left[\frac{E_V - E_F}{K_B T}\right]$$

$$\left[m_e^*\right]^{3/2} \exp\left[\frac{E_F - E_C}{K_B T}\right] = \left[m_h^*\right]^{3/2} \exp\left[\frac{E_V - E_F}{K_B T}\right]$$

$$\left[\frac{m_h^*}{m_e^*}\right]^{3/2} = \exp\left[\frac{E_F - E_C - E_V + E_F}{K_B T}\right]$$

$$\left[\frac{m_h^*}{m_e^*}\right]^{3/2} = \exp\left[\frac{2E_F - E_C - E_V}{K_B T}\right]$$

i.e., Density of electrons (N_e) = Density of holes (N_h)

Taking log on both sides, we get:

$$\log\left[\frac{m_h^*}{m_e^*}\right]^{3/2} = \frac{[2E_F - E_C - E_V]}{K_B T}$$

$$\frac{3}{2}\log\left[\frac{m_h^*}{m_e^*}\right] = \frac{[2E_F - E_C - E_V]}{K_B T}$$

$$\frac{3}{2}K_B T\log\left[\frac{m_h^*}{m_e^*}\right] = [2E_F - (E_C + E_V)]$$

$$2E_F = (E_C + E_V) + \frac{3}{2}K_B T\log\left[\frac{m_h^*}{m_e^*}\right]$$

$$E_F = \frac{E_C + E_V}{2} + \frac{3}{4}K_B T\log\left[\frac{m_h^*}{m_e^*}\right]$$

If $m_h^* = m_e^*$, then at $T = 0K$,

$$= \frac{3}{2}K_B T\log\left[\frac{m_h^*}{m_e^*}\right]$$

$$= E_F = \left[\frac{E_C + E_V}{2}\right]$$

Therefore, the Fermi level lies in the midway between conduction level E_c and valence level E_v at T = 0 K.

But in general, $m_h^* = m_e^*$ so the fermi level is a function of temperature and is raised slightly with the temperature.

4.2 Conductors, Semiconductors and Insulators

By considering the electrical properties of materials, we can divide them into three groups such as:

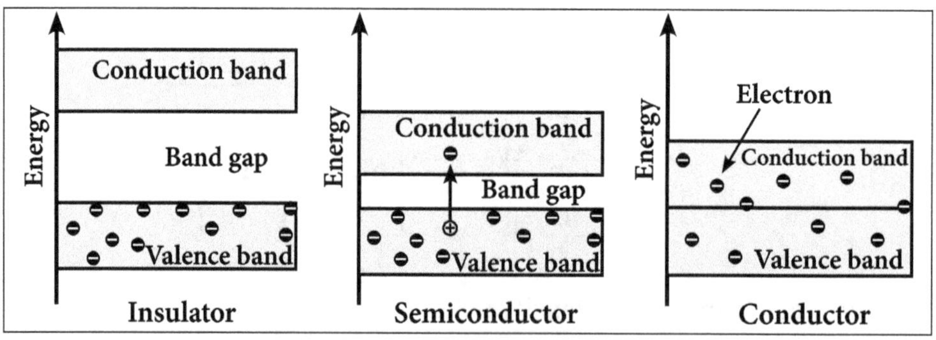

- Conductors: Materials that permit the flow of electrons are called conductors. Conductors have lot of free electrons. Conductors are materials with high conductivities, greater than 10^3 S/cm. Example, all metals and semi-metals like carbon, graphite, antimony, arsenic, gold, silver, copper, etc.
- Insulators: Materials that blocks the flow of electrons are known as insulators. Insulators have very few free electrons. Insulators have an electrical conductivity less than 10^{-8} S/cm. Example, rubber, glass, Teflon, mica, plastic, glass, wood, etc.

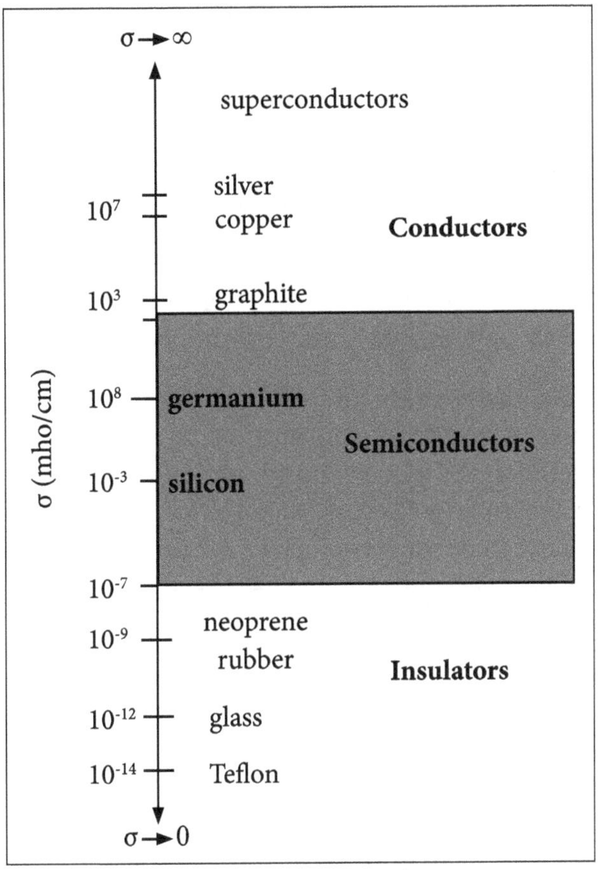

Semiconductors and its conductivity.

- Semiconductors: Materials whose electrical conductivity falls between those of conductors and insulators are termed as semiconductors. Semiconductors are part time conductors whose conductivity can be controlled. These materials lie between the extremes of good conductors and good insulators. They have a electrical conductivity between 10^{-8} S/cm and 10^3 S/cm.

They are crystalline materials which are insulators when pure and will become as conductor when impurity is added or in response to light, heat, voltage, etc. Examples, elements like germanium, silicon, selenium and compounds like gallium arsenide and indium antimonide.

4.2.1 Semiconductors: Intrinsic and Extrinsic

Intrinsic Semiconductors

Semiconductor is a material whose conductivity lies in between that of the conductors and insulators. Semiconductors which are chemically pure, meaning free of impurities, are called Intrinsic Semiconductors or Un-doped Semiconductor or i-type Semiconductor.

The most common intrinsic semiconductors are Silicon and Germanium, which belong to Group IV of the periodic table. The atomic numbers of Silicon and Germanium are 14 and 32, which yields their electronic configuration as $1s^2\ 2s^2\ 2p6\ 3s^2\ 3p^2$ and $1s^2\ 2s^2\ 2p^6\ 3s^2\ 3p^6\ 4s^2\ 3d^{10}\ 4p^2$, respectively. This indicates that both Si and Ge have four electrons each in their outer-most i.e. valence shell. These electrons are called valence electrons and are responsible for the conduction-properties of the semiconductors.

Crystal lattice of Silicon (it is the same even for Germanium) in two-dimension is as shown in figure (1). Here, it is seen that each valence electron of a Si atom pairs with the valence electron of the adjacent Si atom to form covalent bond.

After pairing, the intrinsic semiconductor will be deprived of free charge carriers which are nothing but the valence electrons. Hence, at 0 K the valence band will be full of electrons while the conduction band will be empty. At this stage, no electron in the valence band would gain sufficient energy to cross the forbidden energy gap of the semiconductor material. Thus, the intrinsic semiconductors act as insulators at 0 K.

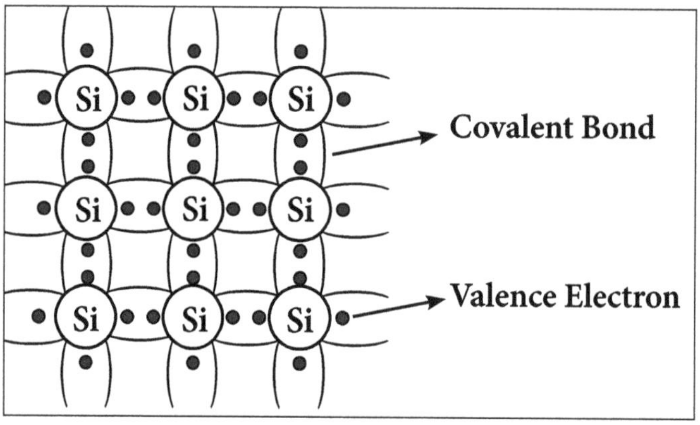

(1) 2-D Crystal Lattice Silicon.

However, at room temperature, the thermal energy may cause a few of the covalent bonds to break, thus, generating free electrons. Thus, the electrons generated get excited and move into the conduction band from the valence band, overcoming the energy barrier. During this process, each electron leaves behind a hole in the valence band. The electrons and holes created in this way are called intrinsic charge carriers and are responsible for the conductive properties exhibited by the intrinsic semiconductor material.

Although, the intrinsic semiconductors are capable of conducting at room temperature, it is to be noted that the conductivity so exhibited is low as there are only a few charge carriers. But as the temperature increases, more and more covalent bonds break which results in more and more number of free electrons.

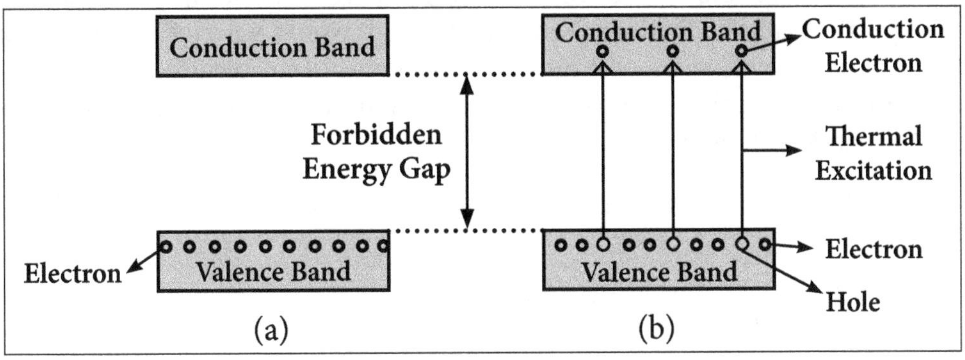

(2) Energy band diagram of intrinsic semiconductor at:
(a) 0 K and (b) Temperature > 0 K.

This in turn results in the movement of greater number of electrons into the conduction band from the valence band. As the population of the electrons in the conduction band increases, the conductivity of the intrinsic semiconductor also increases. However, the number of electrons (n_i) in the intrinsic semiconductor always remains equal to the number of holes in it (p_i).

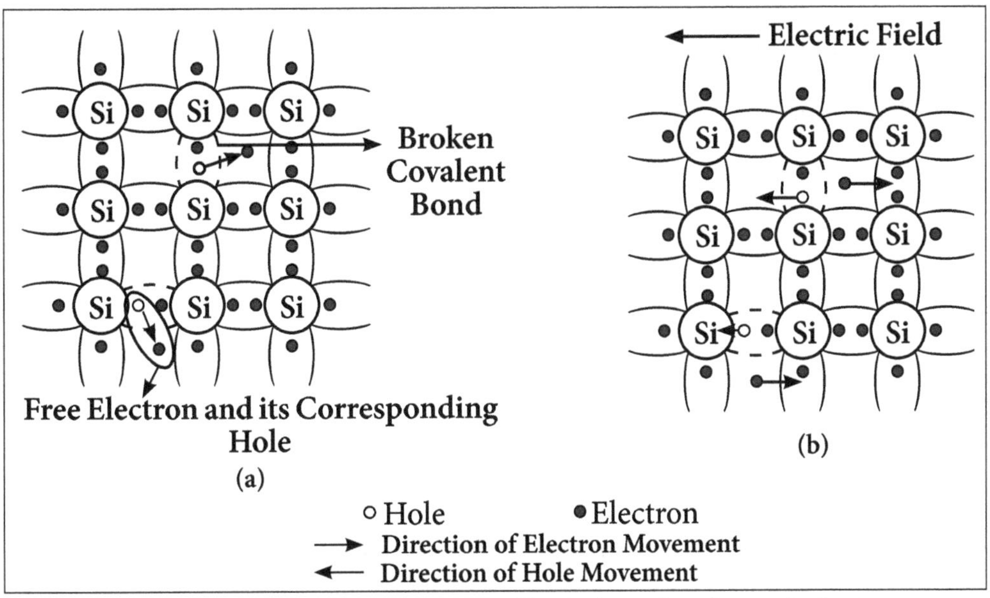

(3) Conduction mechanism in case of intrinsic semiconductors in the:
(a) Absence of Electric field and (b) Presence of Electric field.

On applying the electric field to such an intrinsic semiconductor, the electron-hole pairs can be made to drift under its influence. In this case, the electrons move in the direction opposite to that of the applied field while the holes move in the direction of the electric field as shown by figure (3b). This means that the direction along which the electrons and holes move are mutually opposite.

This is because, as an electron of a particular atom moves towards say, left, by leaving a hole in its place, the electron from the neighbouring atom occupies its place by re-combining with that hole. However, while doing so, it would have left one more hole in its place. This can be viewed as the movement of holes in the semiconductor material. These two movements, although opposite in direction, result in the total flow of current through the semiconductor.

Mathematically the charge carrier densities in intrinsic semiconductors are given by:

$$n_i = N_c e\left(\frac{E_F - E_C}{kT}\right) \text{ where, } N_C = 2\left(\frac{2\pi m_e kT}{h^2}\right)^{3/2}$$

$$p_i = N_v e\left(\frac{E_v - E_F}{kT}\right) \text{ where, } N_v = 2\left(\frac{2\pi m_h kT}{h^2}\right)^{3/2}$$

Where,

N_c - The effective densities of states in the conduction band.

N_v - The effective densities of states in the valence band.

k - The Boltzmann constant (1.38×10^{-23}) JK^{-1}

T - The temperature.

E_F - The Fermi energy.

E_v - The level of valence band.

E_c - The level of conduction band.

h - The Planck constant (6.624×10^{-24}) Js.

m_h - The effective mass of a hole.

m_e - The effective mass of an electron.

Carrier Concentration in Intrinsic Semiconductors

In a semiconductor, both electrons and holes are charge carriers (known as carrier concentration). A semiconductor in which holes and electrons are created by thermal excitation across the energy gap is called as an intrinsic semiconductor.

In an intrinsic semiconductor, the number of holes is equal to the number of free electrons.

At T = 0 K, valence band is completely filled and conduction band is completely empty. Thus, the intrinsic semiconductor behaves as a perfect insulator.

At T > 0 K, the electron from the valence band shifted to the conduction band across the band gap.

Thus, there are number of free electrons and holes in intrinsic semiconductor. Fermi level lies in the midway between the conduction band and valance band in the intrinsic semiconductors.

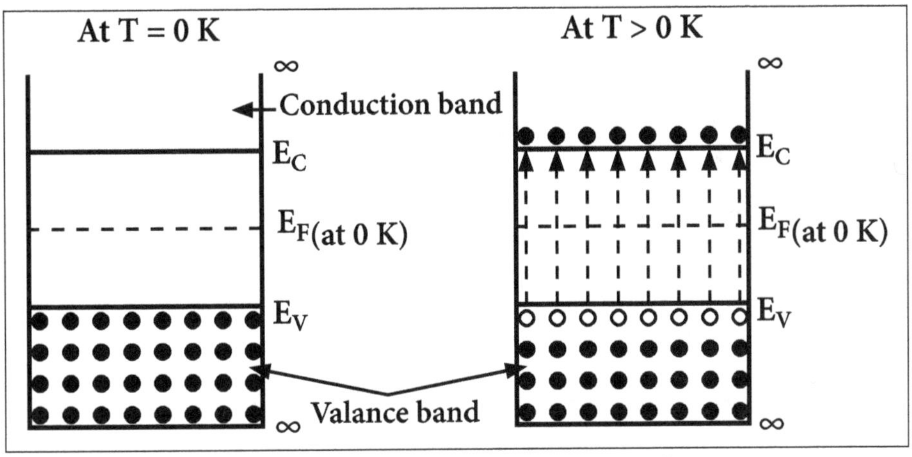

Energy Band Diagram in an Intrinsic Semiconductor.

Density of electrons in conduction band is given by:

$$N_e = 2\left[\frac{\left(2\pi m_e^* K_B T\right)}{h^2}\right]^{3/2} \exp\left[\frac{E_F - E_C}{K_B T}\right]$$

Density of holes in valence band is given by:

$$N_h = 2\left[\frac{\left(2\pi m_h^* K_B T\right)}{h^2}\right]^{3/2} \exp\left[\frac{E_V - E_F}{K_B T}\right]$$

Intrinsic carrier concentration:

In intrinsic semiconductors $N_e = N_h = n_i$ is called intrinsic carrier concentration.

$$n_i^2 = N_e N_h$$

$$n \quad 2\left[\frac{2\pi m_e K_B T}{h^2}\right]^{3/2} \exp\left[\frac{E_F - E_C}{K_B T}\right] 2\left[\frac{2\ m_h K_B T}{h^2}\right]^{3/2} \exp\left[\frac{E_V - E_F}{K_B T}\right]$$

$$n_i^2 = 4\left[\left[\frac{2\pi K_B T}{h^2}\right]^{3/2}\right]^2 \left[m_e^* m_h^*\right]^{3/2} \exp\left[\frac{-[E_C - E_V]}{K_B T}\right]$$

$$n_i^2 = 4\left[\left[\frac{2\pi K_B T}{h^2}\right]^{3/2}\right]^2 \left[m_e^* m_h^*\right]^{3/2} \exp\left[\frac{-E_g}{K_B T}\right]$$

$$n_i = \left[4\left[\left[\frac{2\pi K_B T}{h^2}\right]^{3/2}\right]^2 \left[m_e^* m_h^*\right]^{3/2} \exp\left[\frac{-E_g}{K_B T}\right]\right]^{1/2}$$

$$n_i = 2\left(\frac{2\pi K_B T}{h^2}\right)^{3/2} \left[m_e^* m_h^*\right]^{3/4} \exp\left(\frac{-E_g}{K_B T}\right)^{1/2}$$

$$n_i = 2\left[\frac{2\pi K_B T}{h^2}\right]^{3/2} \left[m_e^* m_h^*\right]^{3/4} \exp\left[\frac{-E_g}{2K_B T}\right]$$

The above equation is called as intrinsic carrier concentration.

Where, $E_g = E_c - E_v$ is the energy gap between conduction band and valence band.

Therefore for intrinsic semiconductor, even if impurity is added to increase N_e, there will be decrease in N_h and hence the product $N_e N_h$ will remain constant. This is called Law of mass action.

Expression for Electrical Conductivity in Intrinsic Semiconductor

The general expression for the electrical conductivity, $\sigma = n e \mu$

The intrinsic electrical conductivity, $\sigma_i = [n e \mu e + p e \mu_h]$

But, $n = p = n_i$

$$\therefore \sigma_i = [n_i e \mu_e + n_i e \mu_h]$$

$$\sigma_i = n_i e[\mu_e + \mu_h]$$

Where,

μ_e - Electron mobility.

μ_h - Hole mobility.

$$\sigma_i = [\mu_e + \mu_h] e . 2\left[\frac{2\pi K_B T}{h^2}\right]^{3/2} \left[m_e^* m_h^*\right]^{3/4} \exp\left[-\frac{E_g}{2K_B T}\right]$$

The electrical conductivity depends on the negative exponential of band gap E_g between the valance band and conduction band and also for the mobilities of both holes and electrons. The mobilities in a pure semiconductor are determined by the interaction of electron with lattice waves or phonons.

So that we can neglect $[\mu_e + \mu_h]$.

The electrical conductivity is given by:

$$\sigma_i = C\exp\left[-\frac{E_g}{2K_BT}\right] \quad ...(1)$$

Where, C is a constant.

Taking log on both sides of equation (1):

$$\log\sigma_i = \text{Log}\left[C\ \exp\left(-\frac{E_g}{2K_BT}\right)\right]$$

$$\log\sigma_i = \text{Log}\,C + \left(\frac{-E_g}{2K_BT}\right)$$

$$\log\sigma_i = \text{Log}\,C - \left(\frac{E_g}{2K_BT}\right)$$

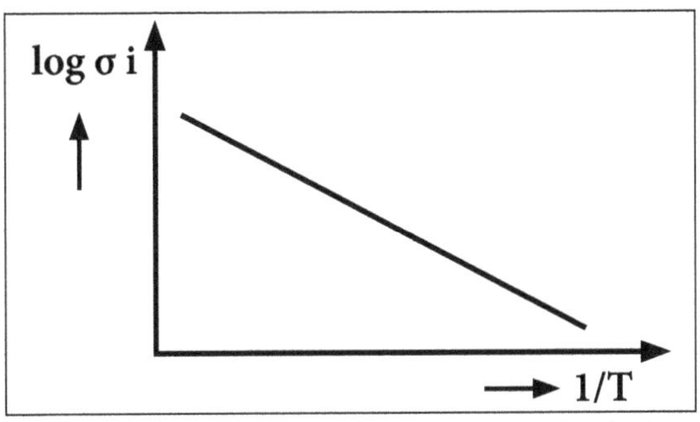

Variation of Electrical conductivity with temperature in intrinsic semiconductor.

A graph is drawn between 1/T and Log σ_i. From the graph it is noted that this electrical conductivity increases with temperature.

Band Gap Determination

We know that the electrical conductivity:

$$\sigma_i = C\exp\left[\frac{-E}{2K_BT}\right]$$

The resistivity is:

$$\rho_i = \frac{1}{\sigma_i}$$

$$\rho_i = \frac{1}{C}\exp\left(\frac{E_g}{2K_B T}\right) \quad ...(2)$$

We know resistivity is the resistance per unit area per unit length:

$$\rho_i = \frac{R_i A}{L} \quad ...(3)$$

Where,

ρ_i - Resistivity.

A - Cross-sectional area.

L - Length.

Equating (2) and (3),

$$\frac{R_i A}{L} = \frac{1}{C}\exp\left[\frac{E_g}{2K_B T}\right]$$

$$R_i = \frac{L}{AC}\exp\left[\frac{E_g}{2K_B T}\right]$$

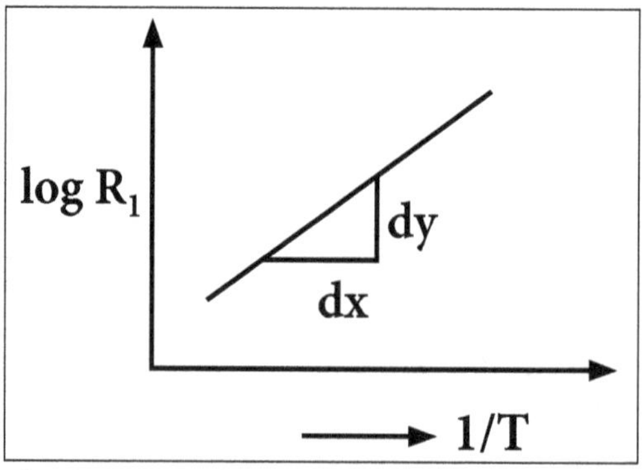

Variation of resistance with temperature in intrinsic semiconductor.

Taking log on both sides:

$$\log R_i = \log\frac{L}{AC} + \left[\frac{E_g}{2K_B T}\right]$$

The above equation gives us a method of determining the energy gap of an intrinsic material. If we find the resistance of the intrinsic semiconductor using carry Foster's bridge at various temperatures, we can plot a graph between 1/T and log R_i.

From the curve:

$$\frac{dy}{dx} = \frac{E_g}{2K_B T}$$

Therefore, by finding the slope of line, we can calculate the energy band gap with the following expression:

$$E_g = 2K_B \left(\frac{dy}{dx} \right) \text{Joules}$$

Extrinsic Semiconductor

A semiconductor to which an impurity at controlled rate is added to make it conductive is known as an extrinsic Semiconductor.

The compound semiconductor is a semiconductor compound composed of elements from two or more different groups of the periodic table. They are also known as direct band gap semiconductors.

i.e., III - V group, II - VI group and IV - VI group.

Here, the recombination of electron and hole takes place directly, during recombination photons are emitted.

Example: GaAs, GaP.

It is well known that the pure semiconductors like Silicon (Si) or Germanium (Ge) are tetravalent (have four electrons in their valence shell) as they belong to Group IV of the periodic table. Thus, if one needs to increase the number of electrons in them, they are to be doped with the Group V elements of the periodic table viz., Phosphorous (P), Arsenic (As), Antimony (Sb), Bismuth (Bi) or Lithium (Li).

This is because Group V elements are pentavalent in nature, meaning which they have five electrons in their valence shell. This indicates that even after the formation of four covalent bonds due to the sharing of four electrons between the Group V and Group IV elements, one more electron (fifth electron of Group V element) would be left-behind as shown by figure (1a).

In other words, pentavalent impurities are said to donate their electron to the tetravalent (pure) semiconductors and are hence called donors. The electron so donated will be very loosely bound to its parent atom and can thus be made 'free' by supplying very minimal amount of thermal energy. Such an excitation causes the electron to transit from its current energy state, donor level ED to the conduction band.

From then on it can actively contribute to the process of conduction in the semiconductor material along with the other free electrons generated due to the breaking of covalent bonds. From this, the total number of electrons in such a material is the sum of thermally generated electrons and the electrons donated by the donor atoms.

However, the number of holes in the same material remains equal to the number of thermally generated electrons only. This indicates that the number of electrons in this kind of material exceeds the number of holes, which means the electrons will be the majority charge carriers while the holes will be the minority carriers. Thus, these materials are named as n-type semiconductors.

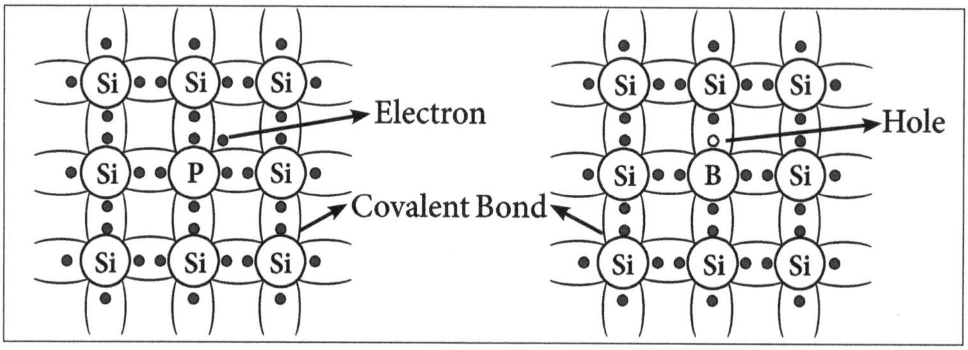

(1) 2-D Crystal lattice of silicon doped with (a) Phosphorous, (b) Boron.

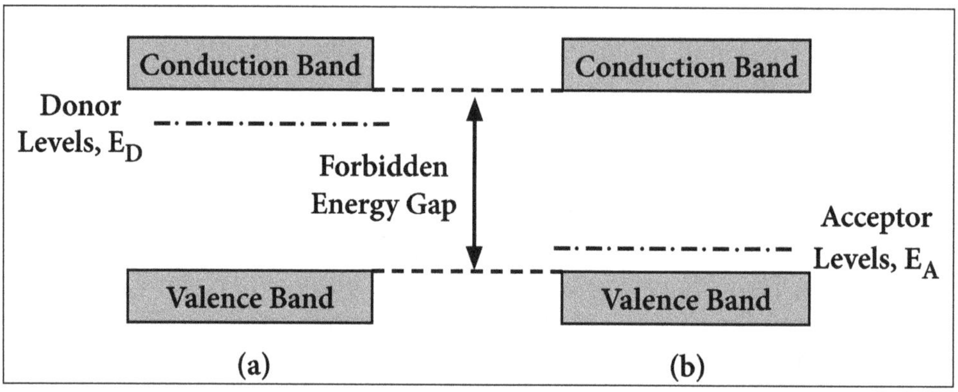

(2) Energy band diagram of (a) n-type Extrinsic semiconductor, (b) p-type Extrinsic semiconductor.

Similarly, the Group IV elements can also be doped with Group III elements of the periodic table like Boron (B), Aluminium (Al), Nitrogen (N), Gallium (Ga) or Indium (In). In this case, the resultant material will have three covalent bonds formed by the pairing-up of three electrons from each atom. However, due to the trivalent (three valence electrons) nature of Group III elements, there will be one insufficient electron to form the fourth covalent bond. This deficiency in electron signifies nothing but the hole as shown in figure (1b).

The number of holes so formed is equal to the number of trivalent impurity atoms added into the semiconductor. Each of these holes signifies a vacant space which can readily accept an electron into it and is present at the energy level EA shown in figure (2b). Further due to the nature of 'acceptance' exhibited by the holes induced due to the doping of Group III elements, such elements are referred to as acceptors.

Moreover, as an electron moves into the hole to recombine, it leaves a new hole in its previous place which in turn will be filled by some other electron. This indicates that the movement of electrons in one direction can be viewed as the movement of holes in the opposite direction. Thus, such a semiconductor material becomes conductive in nature.

However, for this case, the total number of holes will be equal to the sum of the holes induced due to doping and the holes generated due to the thermal excitation process, unlike the electrons which are produced only due to the process of thermal excitation. This means that here the number of holes exceeds the number of electrons for which the holes will be the majority charge carriers while the electrons will be the minority charge carriers. Due to these kinds of extrinsic semiconductor materials are called as p-type semiconductors.

Carrier Concentration in p-Type Semiconductor

p-Type Semiconductor: If trivalent (Aluminum, Gallium, Indium) impurities are doped with pure semiconducting material the holes are produced, this is called p-Type semiconductor.

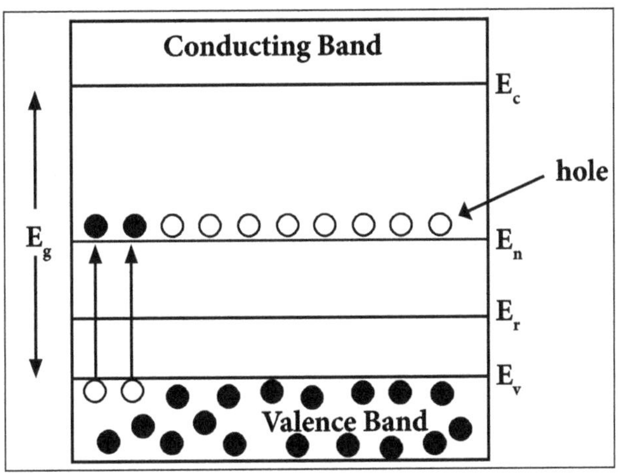

Energy level diagram for P-type semiconductor.

We know that:

Density of holes in the valence band in an intrinsic semiconductors is given by:

$$N_h = 2\left[\frac{2\pi m_h^* K_B T}{h^2}\right]^{3/2} \exp\left[\frac{E_V - E_F}{K_B T}\right] \quad ...(4)$$

Put,

$$N_V = 2\left[\frac{2\pi m_h^* K_B T}{h^2}\right]^{3/2} \quad ...(5)$$

Density of holes is:

$$N_h = N_V \exp\left[\frac{E_V - E_F}{K_B T}\right] \quad ...(6)$$

Density of ionized acceptor atoms is,

$$N_A F[E_A] = N_A \exp\left[\frac{E_F - E_A}{K_B T}\right] \quad ...(7)$$

At equilibrium condition:

[Number of holes per unit volume in valence band equation (6)] = [Number of electron per unit volume in acceptor energy level equation (7)].

$$N_V \exp\left[\frac{E_V - E_F}{K_B T}\right] = N_A \exp\left[\frac{E_F - E_A}{K_B T}\right]$$

$$\exp\left[\frac{E_V - E_F}{K_B T}\right]\exp\left[\frac{-[E_F - E_A]}{K_B T}\right] = \frac{N_A}{N_V}$$

$$\exp\left[\frac{E_V - E_F - E_F + E_A}{K_B T}\right] = \frac{N_A}{N_V}$$

Taking log on both sides, we get:

$$\log\left[\exp\left[\frac{E_V - E_F - E_F + E_A}{K_B T}\right]\right] = \log\left[\frac{N_A}{N_V}\right]$$

$$\left[\frac{E_V - E_F - E_F + E_A}{K_B T}\right] = \log\left[\frac{N_A}{N_V}\right]$$

$$E_V - E_F - E_F + E_A = K_B T \log\left[\frac{N_A}{N_V}\right]$$

$$-2E_F = -[E_V + E_A] + K_B T \log\left[\frac{N_A}{N_V}\right]$$

$$E_F = \frac{(E_V + E_A)}{2} - \frac{K_B T}{2}\log\left[\frac{N_A}{N_V}\right]$$

At T = 0 K,

$$E_F = \frac{(E_V + E_A)}{2} - \frac{K_B T}{2}\log\left[\frac{N_A}{2\left(\frac{2\pi m_h^* K_B T}{h^2}\right)}\right] \qquad ...(8)$$

At 0 K, fermi level in p-type semiconductor lies exactly at the middle of the acceptor level and the top of the valance band.

$$E_F = \left[\frac{(E_V + E_A)}{2}\right]$$

Expression for the Density of Holes in Valence band in Terms of N_A

As the temperature is increased more and more, the acceptor atoms are ionized. Further increase in temperature results in generation of electron hole pairs due to breaking of covalent bonds and materials tends to behave in an intrinsic manner. The fermi level gradually moves towards the intrinsic fermi level.

We know density of holes in valence band is given by:

$$N_h = 2\left[\frac{2\pi m_h^* K_B T}{h^2}\right]^{3/2} \exp\left[\left(\frac{E_V - E_F}{K_B T}\right)\right] \qquad ...(9)$$

Substituting the equation (8) in (9) we get:

$$N_h = 2\left(\frac{2\pi m_h^* K_B T}{h^2}\right)^{\frac{3}{2}} \exp - \frac{K_B T}{2}\log\left[\frac{E_V - \left[\left(\frac{(E_V + E_A)}{2}\right) - \frac{K_B T}{2}\log\frac{N_A}{2\left(\frac{2\pi m_h^* K_B T}{h^2}\right)^{\frac{3}{2}}}\right]}{K_B T}\right]$$

$$=2\left[\frac{\left(2\pi m_h^* K_B T\right)}{h^2}\right]^{3/2}\exp\left[\left(\frac{E_V-E_A}{2K_BT}\right)+\frac{1}{2}\log\left(\frac{N_A}{2\left[\frac{\left[2\pi m_h^* K_B T\right]}{h^2}\right]^{3/2}}\right)\right]$$

$$=2\left[\frac{\left(2\pi m_h^* K_B T\right)}{h^2}\right]^{3/2}\exp\left[\left(\frac{E_V-E_A}{2K_BT}\right)+\log\left(\frac{N_A}{2\left[\frac{\left[2\pi m_h^* K_B T\right]}{h^2}\right]^{3/2}}\right)^{1/2}\right]$$

$$=2\left[\frac{\left(2\pi m_h^* K_B T\right)}{h^2}\right]^{3/2}\exp\left[\left(\frac{EV-EA}{2K_BT}\right)+\log\left(\frac{N_A{}^{1/2}}{2^{1/2}\left[\frac{\left[2\pi m_h^* K_B T\right]}{h^2}\right]^{3/4}}\right)\right]$$

$$=2\left[\frac{\left(2\pi m_h^* K_B T\right)}{h^2}\right]^{3/2}\exp\left(\frac{E_V-E_A}{2K_BT}\right)\left(\frac{N_A{}^{1/2}}{2^{1/2}\left[\frac{\left[2\pi m_h^* K_B T\right]}{h^2}\right]^{3/4}}\right)$$

$$=\left(2N_A\right)^{1/2}\exp\left[\frac{E_V-E_A}{2K_BT}\right]\left[\frac{\left(2\pi m_h^* K_B T\right)}{h^2}\right]^{3/2}\left[\frac{\left(2\pi m_h^* K_B T\right)}{h^2}\right]^{-3/4}$$

$$=\left(2N_A\right)^{1/2}\exp\left[\frac{E_V-E_A}{2K_BT}\right]\left[\frac{\left(2\pi m_h^* K_B T\right)}{h^2}\right]^{3/4}$$

Here, $E_A - E_V = \Delta E$ is known as ionization energy of acceptors.

i.e., ΔE represents the energy required for an electron to move from valance band (E_V) to acceptor energy level (E_A).

$$N_h=\left(2N_A\right)^{1/2}\exp\left[\frac{-\Delta E}{2K_BT}\right]\left[\frac{\left(2\pi m_h^* K_B T\right)}{h^2}\right]^{3/4}$$

Carrier Concentration in n-type Semiconductor

If pentavalent (Phosphorous, Arsenic, Antimony) impurities are doped with pure semi-conducting material the free electrons are produced, this is called n-type semiconductor.

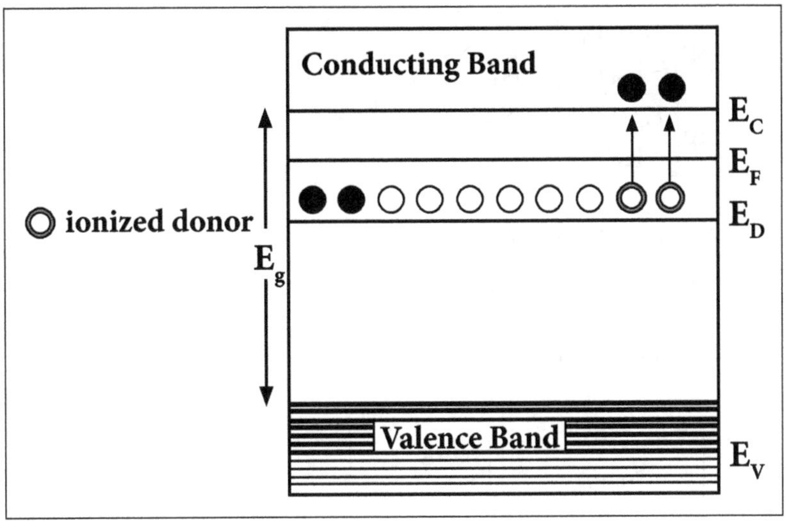

Energy level diagram for n-type semiconductor.

We know that,

Density of electrons in conduction band in an intrinsic semiconductor is given by:

$$N_e = \left[2\left[\frac{2\pi m_e^* K_B T}{h^2}\right]^{3/2} \exp\left[\frac{E_F - E_C}{K_B T}\right]\right] \quad \text{...(10)}$$

Put,

$$N_C = 2\left[\frac{2\pi m_e^* K_B T}{h^2}\right]^{3/2}$$

Density of electrons is:

$$N_e = N_C \exp\left[\frac{E_F - E_C}{K_B T}\right] \quad \text{...(11)}$$

Density of ionized donor atoms is given by:

$$N_D\left[1 - F(E_D)\right] = N_D \exp\left[\frac{E_D - E_F}{K_B T}\right] \quad \text{...(12)}$$

At equilibrium condition,

[Number of electrons per unit volume in conduction band equation (11)] = [Number of holes per unit volume in donor energy level equation (12)]:

$$N_C \exp\left[\frac{E_F - E_C}{K_B T}\right] = N_D \exp\left[\frac{E_D - E_F}{K_B T}\right]$$

$$\frac{\exp\left[\dfrac{E_F - E_C}{K_B T}\right]}{\exp\left[\dfrac{E_D - E_F}{K_B T}\right]} = \frac{N_D}{N_C}$$

$$\exp\left[\frac{E_F - E_C - E_D + E_F}{K_B T}\right] = \frac{N_D}{N_C}$$

Taking log on both sides, we get:

$$\log\left[\exp\left[\frac{E_F - E_C - E_D + E_F}{K_B T}\right]\right] = \log\left[\frac{N_D}{N_C}\right]$$

$$\left[\frac{E_F - E_C - E_D + E_F}{K_B T}\right] = \log\left[\frac{N_D}{N_C}\right]$$

$$2E_F = (E_C + E_D) + K_B T \log\left[\frac{N_D}{N_C}\right]$$

$$E_F = \frac{(E_C + E_D)}{2} + \frac{K_B T}{2} \log\left[\frac{N_D}{N_C}\right] \qquad ...(13)$$

At T = oK,

$$E_F = \left[\frac{(E_C + E_D)}{2}\right] \qquad ...(14)$$

Thus, the fermi level in N-type semiconductor lies exactly in middle of the conduction level (E_C) and donor level (E_D).

This equation shows that the electron concentration in the conduction band is proportional to the square root of the donor concentration.

Expression for the Density of Electrons in Conduction Band in Terms of N_D

As the temperature is increased more and more, the donor atoms are ionized and the fermi level drops. For a particular temperature all donor atoms are ionized, further increase in temperature results in generation of electron hole pairs due to breaking of covalent bonds and materials tends to behave in an intrinsic manner.

We know density of electrons in conduction band is given by:

$$N_e = 2\left[\frac{(2\pi m_e^* K_B T)}{h^2}\right]^{3/2} \exp\left[\frac{E_F - E_C}{K_B T}\right] \quad ...(15)$$

Substituting the equation (13) in (15) we get:

$$N_e = 2\left[\frac{(2\pi m_e^* K_B T)}{h^2}\right]^{3/2} \exp\left[\frac{\left(\frac{E_C + E_D}{2}\right) + \frac{K_B T}{2}\log\left(\frac{N_D}{2\left[\frac{[2\pi m_e^* K_B T]}{h^2}\right]^{3/2}}\right) - E_C}{K_B T}\right]$$

$$= 2\left[\frac{(2\pi m_e^* K_B T)}{h^2}\right]^{3/2} \exp\left[\left(\frac{E_D - E_C}{2K_B T}\right) + \log\left(\frac{N_D}{2\left[\frac{[2\pi m_e^* K_B T]}{h^2}\right]^{3/2}}\right)^{1/2}\right]$$

$$= 2\left[\frac{(2\pi m_e^* K_B T)}{h^2}\right]^{3/2} \exp\left[\left(\frac{E_D - E_C}{2K_B T}\right) + \log\left(\frac{N_D^{1/2}}{2^{1/2}\left[\frac{[2\pi m_e^* K_B T]}{h^2}\right]^{3/4}}\right)\right]$$

$$= 2\left[\frac{(2\pi m_e^* K_B T)}{h^2}\right]^{3/2} \exp\left(\frac{E_D - E_C}{2K_B T}\right)\left(\frac{N_D^{1/2}}{2^{1/2}\left[\frac{[2\pi m_e^* K_B T]}{h^2}\right]^{3/4}}\right)$$

$$= (2N_D)^{1/2} \exp\left[\frac{E_D - E_C}{2K_B T}\right]\left[\frac{(2\pi m_e^* K_B T)}{h^2}\right]^{3/2}\left[\frac{(2\pi m_e^* K_B T)}{h^2}\right]^{3/2}\left[\frac{(2\pi m_e^* K_B T)}{h^2}\right]^{-3/4}$$

$$= (2N_D)^{1/2} \exp\left[\frac{E_D - E_C}{2K_B T}\right]\left[\frac{(2\pi m_e^* K_B T)}{h^2}\right]^{3/4}$$

Here, $E_C - E_D = \Delta E$ is known as ionization energy of donors.

i.e., ΔE represents the amount of energy required to transfer an electron from donor energy level (E_D) to conduction band (E_C).

$$N_e = (2N_D)^{1/2} \exp\left[\frac{-\Delta E}{2K_B T}\right]\left[\frac{(2\pi m_e^* K_B T)}{h^2}\right]^{3/4}$$

4.3 Electrical Conduction Phenomenon

The concepts of mobility and scattering are basic ingredients of electrical conduction phenomenon in electronic materials.

Mobility

Mobility is the essential property that characterizes the carrier transport. The carrier mobility (μ) also plays a key role in characterizing the transport co-efficient in semiconductor devices. Mobility is defined as the drift velocity (v_0) per unit electric field and can be written as:

$$\mu = \frac{v_0}{E_0} \quad ...(16)$$

Where,

E_0 is the applied electric field.

Relationship of Mobility to Scattering

The force generated on a carrier of charge e due to the application of the electric field E_0 is eE_0. The application of Newton's second law leads to the result that in a steady state $eE_0 = m^*(v_0/\tau)$ where, r is known as the momentum relaxation time and has the significance that in the absence of an electric field, the momentum reduces to e^{-1} of its initial value in the time τ sec. Thus, mobility is expressed as:

$$\mu = \frac{e\tau}{m^*} \quad ...(17)$$

In SI units, e is in coulomb, τ is in second and m* is in kg. The dimensional substitution of these quantities in equation (17) leads to the fact that the dimension of mobility is $m^2/(V.s)$. In general, τ and m* are the functions of carrier energy and mobility for the carriers in isotropic energy bands can be expressed as:

$$\mu = e\left\langle \frac{\tau(E)}{m^*(E)} \right\rangle \quad ...(18)$$

Where, $\langle\tau(E)/m^*(E)\rangle$ represents the average value of the bracketed quantity for all the carriers. Mobility is a measure of the ease of carrier motion within a semiconductor crystal. Carrier mobility varies directly with the momentum relaxation time when mass is constant and inversely with the carrier effective mass when the momentum relaxation time is constant. The determination of mobility needs the expressions of τ(E) and m*(E), respectively.

$\tau(E)$ can be determined through the solution of the Boltzmann transport equation under different mechanisms of scattering and the specifications of E-k dispersion relations. $m^*(E)$ can be derived from the specified dispersion relation of the carriers. Thus, (μ) changes from semiconductor to semiconductor and also with respect to doping, temperature and other parameters.

4.3.1 P-type and n-type Semiconductors

p-type Semiconductor

When a small amount of trivalent impurity is added to a pure semiconductor, it is called p-type semiconductor. The trivalent impurity has three valence electrons. These elements are gallium, boron or indium. Such an impurity is called acceptor impurity.

Consider the formation of a p-type material by adding gallium (Ga) into silicon (Si). The gallium atom has three valence electrons. So, gallium atom fits in the silicon crystal in such a way that its three valence electrons form covalent bonds with the three adjacent silicon atoms. Being short of one electron, the fourth covalent bond in the valence shell is incomplete. The resulting vacancy is called a hole. Such a p-type material formation is represented in the below figure. This means that each gallium atom added into silicon atom gives one hole. The number of such holes can be controlled by the amount of impurity added to the silicon.

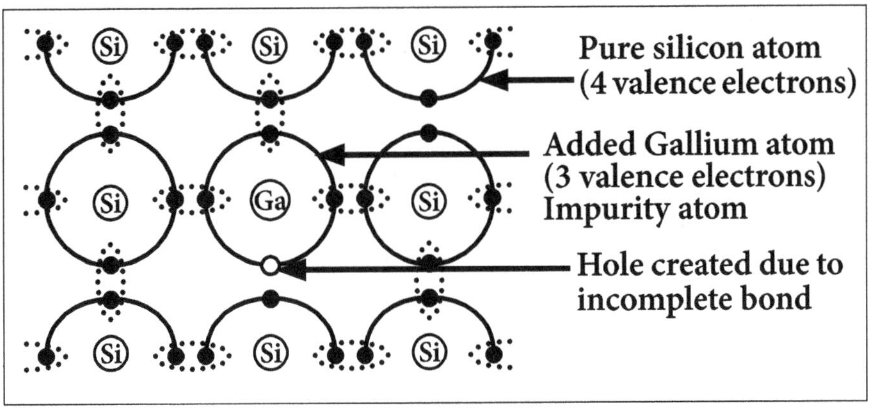

p-type material formation.

As the holes are treated as positively charged, the material is known as p-type material. At room temperature, the thermal energy is sufficient to extract an electron from the neighbouring atom which fills the vacancy in the incomplete bond around the impurity atom, but this creates a vacancy in the adjacent bond from where the electron had jumped, which is nothing but a hole. This indicates that a hole created due to added impurity is ready to accept an electron and hence is called acceptor impurity. Thus, even for a small amount of impurity added, large number of holes gets created in the p-type material.

n-type Semiconductor

When a small amount of pentavalent impurity is added to a pure semiconductor, it is called n-type semiconductor. The pentavalent impurity has five valence electrons. These elements are arsenic, bismuth, phosphorous and antimony. Such an impurity is called donor impurity.

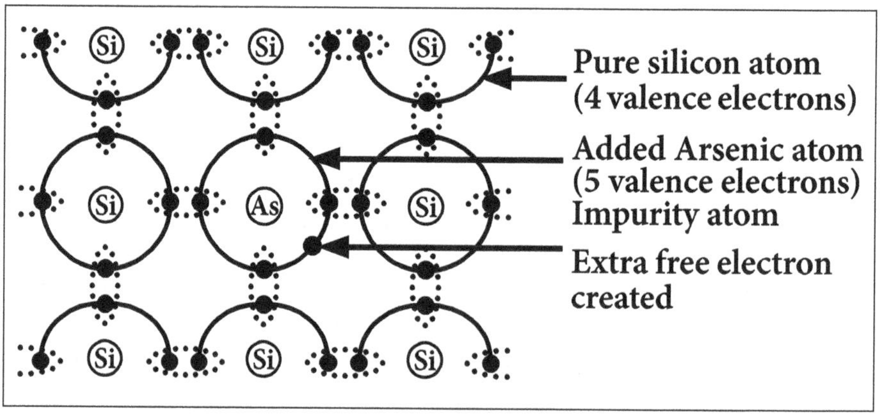

n-type material formation.

Consider the formation of n-type material by adding arsenic (As) into silicon (Si). The arsenic atom has five valence electrons. An arsenic atom fits in the silicon crystal in such a way that its four valence electrons form covalent bonds with four adjacent silicon atoms. The fifth electron has no chance of forming a covalent bond. This spare electron enters the conduction band as a free electron. Such n-type material formation is represented in the above figure.

This means that each arsenic atom added into silicon atom gives one free electron. The number of such free electrons can be controlled by the amount of impurity added to the silicon. Since the free electrons have negative charges, the material is known as n-type material and an impurity donates a free electron hence called donor impurity.

4.3.2 Drift and Diffusion Current

Drift current occurs due to an applied electric field on a semiconductor and diffusion current occurs whenever carrier concentration gradient exists in the material.

Drift Current

Let us consider a semiconductor material of length l at room temperature T Kelvin. At this temperature, free electron and hole concentrations are n and p, respectively. Free electrons and holes possess random motions, so that they do not possess any resultant velocity in any direction inside the crystal. When a voltage V is applied across the semiconductor, the intensity of electric field E in the material is equal to V/l.

This electric field E exerts a force eE on each carrier, so that an acceleration eE/m acts on the carriers, the velocities of electrons and holes are increased in opposite directions up to certain constant value called drift velocity of carriers.

These velocities will not increase to large values because of electrical resistance due to thermal vibrations of ions or atoms at that temperature. The drift velocities of electrons and holes produce drift current in the material.

The drift current density of electrons is given by:

$$J_{n,drift} = nev_{de} \quad ...(19)$$

Where,

v_{de} = Drift velocity of electrons

n = Number of conduction or free electrons per unit volume of the material

The current density of electrons can also be represented using Ohm's law as,

$$J_{n,drift} = \sigma_n E \quad ...(20)$$

Where, σ_n is the electrical conductivity of free electrons.

From equations (19) and (20) we have:

$$\sigma_n E = nev_{de} \quad ...(21)$$

The drift velocity of an electron can be found in the following way:

The moment we apply electric field E on the semiconductor, the drift velocity becomes zero. The electric field exerts a force eE on each electron, so that, each electron acquires constant drift velocity vde after time τ, called collision time. v_{de} can be obtained from equations of motion (v = u + at):

$$v_{de} = 0 + \left(\frac{eE}{m}\right)T = \left(\frac{eT}{m}\right)E \quad ...(22)$$

In equation (22), eτ/m is constant and it can be represented as μ_n = eτ/m, called mobility of electrons. Therefore, equation (22) can be represented as:

$$v_{de} = \mu_n E \quad ...(23)$$

Substituting equation (23) in (21) we have:

$$\sigma_n E = ne\mu_n E$$

Or,

$$\sigma_n = ne\mu_n \quad ...(24)$$

Similarly, the conductivity of holes can be written as:

$$\sigma_p = pe\mu_p \qquad \text{...(25)}$$

The total drift conductivity of the semiconductor is given by:

$$\sigma = \sigma_n + \sigma_p = ne\mu_n + pe\mu_p \qquad \text{...(26)}$$

In an intrinsic semiconductor, $n = p = n_i$

So,

$$\sigma_i = n_i e(\mu_n + \mu_p) \qquad \text{...(27)}$$

The resistivity (ρ) of the semiconductor is given by:

$$\rho = \frac{1}{\sigma} = \left(\frac{1}{ne\mu_n + pe\mu_p}\right) \qquad \text{...(28)}$$

Using equation (26), we can write the drift current densities of electrons and holes as:

$$J_{n,drift} = ne\mu_n E \qquad \text{...(29)}$$

$$J_{p,drift} = pe\mu_p E \qquad \text{...(30)}$$

The total drift current density is equal to the sum of electron and hole drift current densities:

$$J_{drift} = ne\mu_n E + pe\mu_p E \qquad \text{...(31)}$$

The variation of conductivity with temperature can be described as follows:

At 0 K, there are no carriers in the semiconductor. As the temperature of the semiconductor is increased from 0 K to some low temperature, the electrons present in the donor level will go to conduction band in n-type semiconductor or electrons present in valence band will go to acceptor level to produce holes in valence band of p-type semiconductor. By increasing the temperature from 0 K to some low temperature, the carrier concentration increases in the extrinsic semiconductor. Also, the thermal vibrations of ions and hence, electrical resistivity increases.

The increase in conductivity depends on the relative values of the above two quantities, so conductivity increases. As the temperature of extrinsic semiconductor is increased near to room temperature, the carrier concentration would not increase and resistivity increases. So, the materials possess positive temperature coefficient of resistivity.

Above room temperature, the intrinsic carrier concentration and hence, the conductivity increases with rise in temperature compared to the increase in resistivity. So, the material possess negative temperature coefficient of resistivity at and above room temperatures. At higher temperatures, the material exhibits its intrinsic behaviour.

Diffusion Current

Non-uniform concentration of charge carriers produces diffusion current. Non-uniform concentration of charge carriers can be formed by thermal or radiation excitation of a part of the material or by injecting carriers into the material through the surface.

Suppose, the concentration of electrons and holes increases by Δn and Δp at a point in the material due to excitation. These excess carriers diffuse to low concentrated places, the rate of diffusion according to Fick's law is proportional to concentration gradient $\frac{\partial \Delta n}{\partial x}$ along X-direction for excess electrons. As the flow takes place from high concentration region to low concentration region, the rate of flow of electrons is proportional to $-\frac{\partial}{\partial x}\Delta n$ and the rate of flow of electrons through unit area is equal to $-D_n \frac{\partial(\Delta n)}{\partial x}$. Here, D_n is the proportionality constant called diffusion coefficient of electrons. This electron charge flow constitutes electron diffusion current density represented as $J_{n,\,diff}$.

$$J_{n,diff} = (-e) \times \left[-D_n \frac{\partial(\Delta n)}{\partial x}\right] = D_n e \frac{\partial(\Delta n)}{\partial x}$$

Similarly for holes:

$$J_{p,diff} = (e) \times \left[-D_p \frac{\partial \Delta p}{\partial x}\right] = -D_p e \frac{\partial(\Delta p)}{\partial x}$$

Where, D_p is called the diffusion coefficient of holes.

Even though the excess electrons and holes diffuse in the same direction, their current densities are in opposite direction because of opposite charge on electrons and holes.

The total diffusion current is given by:

$$J_{diff} = J_{n,diff} + J_{p,diff}$$

$$= D_n e \frac{\partial(\Delta n)}{\partial x} - D_p e \frac{\partial(\Delta p)}{\partial x}$$

If there is a concentration gradient along the X-direction and if we apply an electric field E on it, the total current in the semiconductor is the sum of drift and diffusion currents. The total current due to electrons and holes are given by:

$$J_n = J_{n,drift} + J_{n,diff}$$

$$= ne\mu_n E + D_n e \frac{\partial(\Delta n)}{\partial x}$$

$$J_p = J_{p,drift} + J_{p,diff}$$

$$= pe\mu_p E - D_p e \frac{\partial(\Delta p)}{\partial x}$$

$$J = J_n + J_p$$

$$J_{total} = ne\mu_n E + D_n e \frac{\partial(\Delta n)}{\partial x} + pe\mu_p E - D_p e \frac{\partial(\Delta p)}{\partial x}$$

Einstein's Equation

Einstein showed the direct relationship between the mobility (μ) and diffusion coefficient (D) of a semiconductor. Consider that a semiconductor is in equilibrium at some temperature T Kelvin with no applied electric field. The distribution of free electrons and holes is uniform and there is no net current flow in any direction.

Any disturbance in carrier concentration leads to diffusion current in the material. As the material is at some temperature T Kelvin, it possesses electrical resistance due to thermal vibrations of ions. The product of resistance and diffusion current results in voltage and electric field in the material. Therefore, in equilibrium, the drift and diffusion currents due to excess concentrations are equal.

So, for electrons, we can write:

$$(\Delta n) e\mu_n E = D_n e \frac{\partial(\Delta n)}{\partial x} \quad ...(32)$$

The force (F) on excess electrons (Δn) to restore equilibrium is given by the product of Δn and electric force eE on each electron. Thus, we have:

$$F = (\Delta n) eE \quad ...(33)$$

Equation (32) can be written as:

$$(\Delta n) eE = \frac{D_n}{\mu_n} e \frac{\partial(\Delta n)}{\partial x} = F \quad ...(34)$$

At temperature T Kelvin, the force on excess electrons to maintain equilibrium depends on the thermal energy of excess electrons [equal to K_BT] times the concentration gradient:

Thus,

$$F = K_B T \frac{\partial(\Delta n)}{\partial x} \qquad ...(35)$$

Equations (34) and (35) are equal.

$$\frac{D_n}{\mu_n} e \frac{\partial(\Delta n)}{\partial x} = K_B T \frac{\partial(\Delta n)}{\partial x}$$

Or,

$$\frac{D_n}{\mu_n} = \frac{K_B T}{e} \qquad ...(36)$$

Or,

$$D_n = \mu_n \frac{K_B T}{e} \qquad ...(37)$$

Similarly, for holes:

$$\frac{D_p}{\mu_p} = \frac{K_B T}{e} \qquad ...(38)$$

Or,

$$D_p = \mu_p \frac{K_B T}{e} \qquad ...(39)$$

From equations (36) and (38), we have:

$$\frac{D_n}{\mu_n} = \frac{D_p}{\mu_p} \qquad ...(40)$$

Or

$$\frac{D_n}{D_p} = \frac{\mu_n}{\mu_p} \qquad ...(41)$$

Equations (36) to (41) are called Einstein's equations. Using the above equations, the diffusion coefficients of electrons and holes can be determined. At 300 K, $\frac{K_B T}{e}$ = 26 mV. The diffusion coefficients for Si are $D_n = 3.6 \times 10^{-3}$ m²/s and $D_p = 1.3 \times 10^{-3}$ m²/s and for Ge, $D_n = 10 \times 10^{-3}$ m²/s and $D_p = 5 \times 10^{-3}$ m²/s.

5

Diodes and Diode Circuits

5.1 Formation of p-n Junction

A diode is formed by joining an n-type semiconductor with a p-type semiconductor. A p-n junction is the interface between n and p regions. Before using the p-n junction as a practical device or as a rectifying device, first we need to bias the junction, i.e., connect a voltage potential across it. On the voltage axis, "Reverse Bias" refers to an external voltage potential which increases the potential barrier.

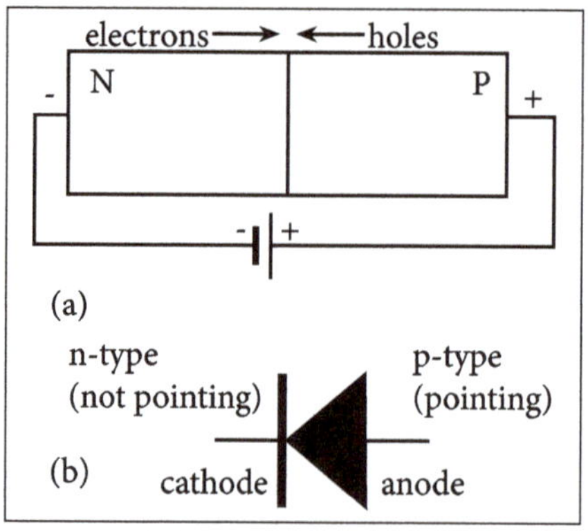

Symbol of p-n junction diode.

- The p region has holes (Majority carriers).
- The n region has free electrons.
- The p and n charge carriers are called as Mobile charges.

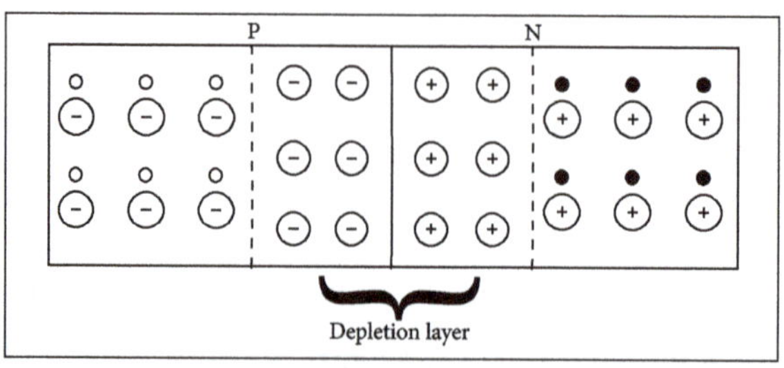

Structure of p-n junction diode.

p-n Junction is formed when some of the holes in p-region and the free electrons in the n-region diffuse into each other and disappear due to recombination.

Since the n-type region has a high electron concentration and p-type has a high hole concentration, electrons diffuse from the n-side to the p-side. Similarly, holes flow by diffusion from the p-side to the n-side.

If the electrons and holes are not charged, this diffusion process would continue till the concentration of electrons and holes on the two sides are the same, this happens if two gases come in contact with each other. However, in a p-n junction, when the electrons and holes move to the other side of the junction, they leave behind exposed charges on the dopant atom sites, which are fixed in the crystal lattice and are unable to move.

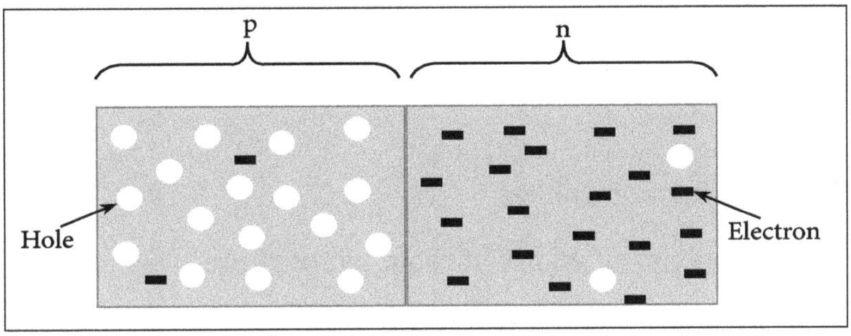

Situation before formation of depletion region.

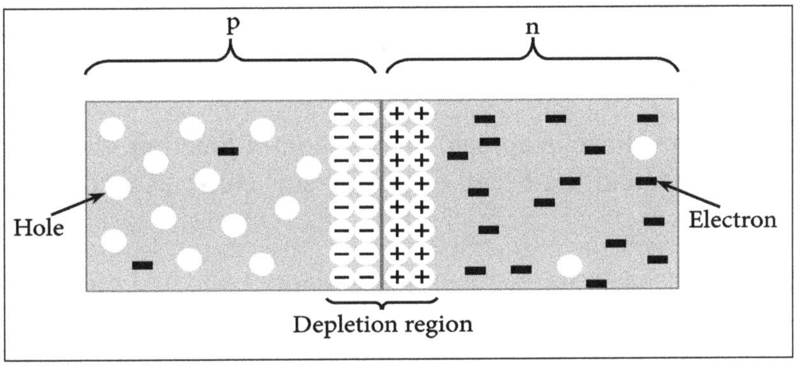

Situation after the formation of depletion region.

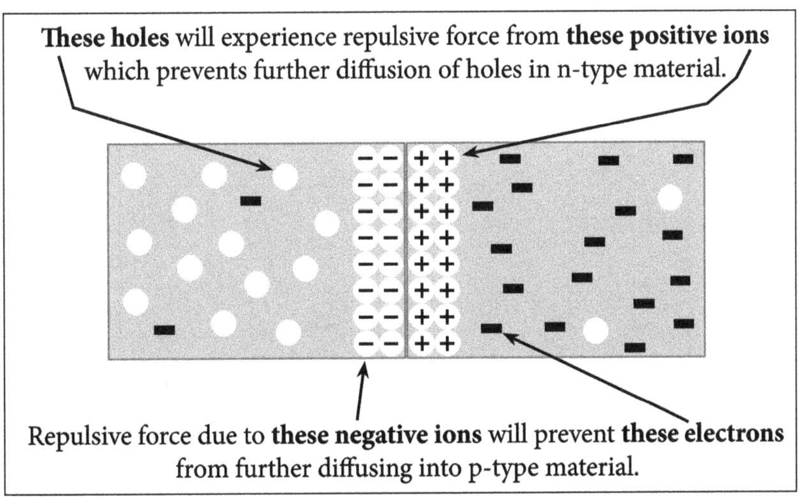

On the n-side, positive ion cores are exposed. On the p-side, negative ion cores are exposed. An electric field Ê forms between the positive ions cores in the n-type material and negative ion cores in the p-type material. This region is known as the depletion region, since the electric field quickly sweeps out the free carriers, hence the region is depleted of free carriers. A built-in potential V_{bi} due to Ê is formed at the junction.

Built-in potential difference across the junction with an open circuit potential is given by:

$$E_o = V_T \, In\left(\frac{N_D.N_A}{n_i^2}\right)$$

Where,

E_o = Zero bias junction voltage.

V_T = Thermal voltage of 26 mV at room temperature.

N_D and N_A = Impurity concentrations.

n_i = Intrinsic concentration.

5.1.1 Built-in Potential

If we assume that no voltage is applied across the p-n junction, then the junction is in thermal equilibrium and the Fermi energy level is constant throughout the entire system.

As we go through the space charge region, the valance band and the conduction band energies must bend, since the relative position of the valance and the conduction bands with respect to the Fermi energy changes between p and n regions.

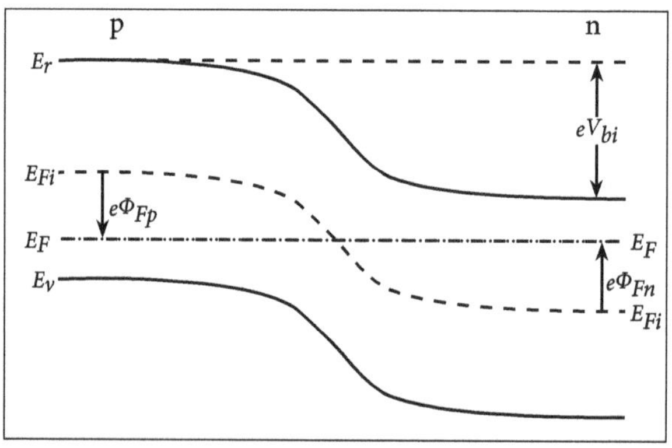

Energy-band diagram of a p-n junction in thermal equilibrium.

Electrons in the conduction band of the n region sees a potential barrier when trying to move into the conduction band of the p region.

This potential barrier is referred as the built-in potential barrier and is given by V_{bi}.

The built-in potential barrier maintains equilibrium between the majority carrier electrons in the n region and the minority carrier electrons in the p region and also between the majority carrier holes in the p region and the minority carrier holes in the n region.

This potential difference across the junction cannot be determined by a voltmeter because new potential barriers will be created between the probes and the semiconductor that will cancel V_{bi}. The potential V_{bi} maintains the equilibrium, so no current is produced by this voltage.

The intrinsic Fermi level is equidistant from the conduction band edge through the junction, thus the built-in potential barrier can be defined as the difference between the intrinsic Fermi level in the p and n regions. We can define the potentials ϕ_{Fn} and ϕ_{Fp} as shown in above figure so we have:

$$V_{bi} = |\phi_{Fn}| + |\phi_{Fp}| \quad ...(1)$$

In the n region, the electron concentration in the conduction band is given by:

$$n_o = N_c \exp\left[-\frac{E_C - E_F}{kT}\right] \quad ...(2)$$

Which can also be written in the form:

$$n_o = n_i \exp\left[\frac{E_F - E_{Fi}}{kT}\right] \quad ...(3)$$

Where n_i and E_{Fi} are the intrinsic carrier concentration and the intrinsic Fermi energy, respectively. We may define the potential ϕ_{Fn} in the n region as:

$$e\phi_{Fn} = E_{Fi} - E_F \quad ...(4)$$

Equation (3) may then be written as:

$$n_o = n_i \exp\left[-\frac{e\phi_{Fn}}{kT}\right] \quad ...(5)$$

Taking the natural logarithm of both sides, putting $n_o = N_d$ and solving for to the potential we obtain:

$$\phi_{Fn} = -\frac{kT}{e}\ln\left(\frac{N_d}{n_i}\right) \quad ...(6)$$

Similarly, in the p region, the hole concentration is given by:

$$p_o = N_a = n_i \exp\left[\frac{E_{Fi} - E_F}{kT}\right] \quad ...(7)$$

Where N_a is the acceptor concentration. We can define the potential ϕ_{Fp} in the p region as:

$$e\phi_{FP} = E_{Fi} - E_F \quad ...(8)$$

Combining equations (7) and (8), we find that:

$$\phi_{Fp} = +\frac{kT}{e}\ln\left(\frac{N_a}{n_i}\right) \quad ...(9)$$

Finally, the built-in potential barrier for the step junction is found by substituting equations (6) and (8) into equation (1), which yields:

$$\text{Built-in Potential, } V_{bi} = \frac{kT}{e}\ln\left(\frac{N_a N_d}{n_i^2}\right) = V_t \ln\left(\frac{N_a N_d}{n_i^2}\right) \quad ...(10)$$

Where,

$V_t = kT/e$ and is defined as the thermal voltage.

N_d and N_a denote donor and acceptor impurity concentrations in the same region, thus forming a compensated semiconductor.

From this point on in the text, N_d and N_a will denote the net donor and acceptor concentrations in the individual n and p regions, respectively. For example if the p region, is a compensated material, then N_a will denote the difference between the actual acceptor and donor impurity concentrations. The parameter N_d is defined in a similar manner for the n region.

5.1.2 Forward and Reverse Biased p-n Junction

Reverse Biased p-n Junction Diode

If the diode is connected in a Reverse Bias condition, a positive voltage is applied to the n-type material whereas a negative voltage is applied to the p-type material.

Thus when the positive voltage is applied to the n-type material, it attracts electrons towards the positive electrode and away from the junction, while the holes in the p-type end are attracted away from the junction towards the negative electrode.

The net result is that the depletion layer grows wider because of a lack of electrons and holes and it presents a high impedance path, almost an insulator. Then the result is that a high potential barrier is created thus preventing current from flowing via the semiconductor material.

Increase in the Depletion Layer due to Reverse Bias

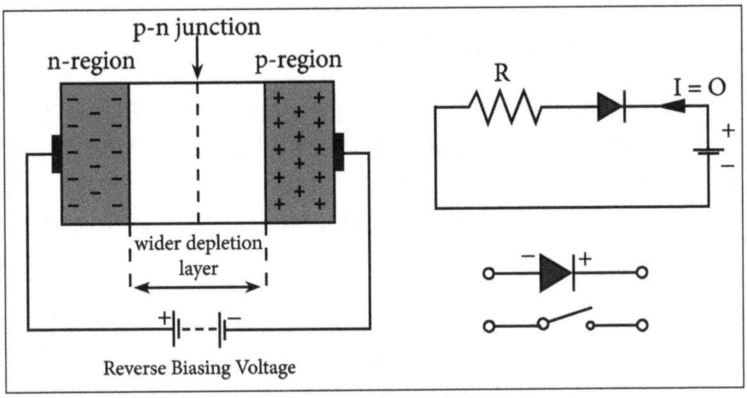

Reverse Bias.

This condition represents a high resistance value to the p-n junction and experimentally zero current flows through the junction diode with an increase in bias voltage. But, a very small leakage current does flow through the junction that can be measured in micro-amperes, (μA).

When the reverse bias voltage V_r is applied to the diode is increased to a sufficiently high enough value, it shall cause the diode's p-n junction to overheat and fail because of the avalanche effect around the junction. This can cause the diode to become shorted and will result in flow of maximum circuit current and this is shown as the step downward slope in the reverse static characteristics curve below.

Reverse Characteristics Curve for a Junction Diode

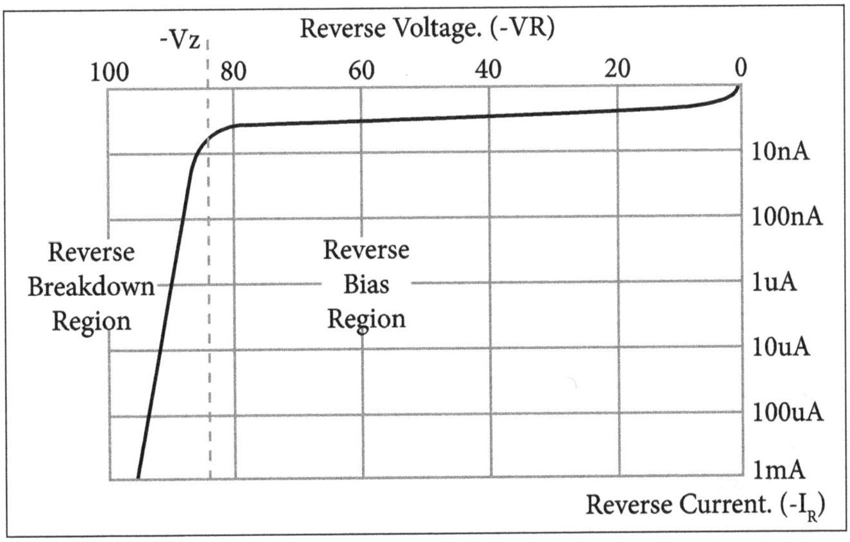

At times, this avalanche effect has practical applications in voltage stabilizing circuits, where a series limiting resistor is used with the diode for limiting this reverse breakdown current to present the maximum value, thereby producing a fixed voltage output across the diode. Such types of diodes are commonly known as zener diodes.

Forward Biased p-n Junction Diode

If the diode is forward biased, a negative voltage is applied to the n-type material and a positive voltage is applied to the p-type material. When this external voltage becomes greater than the value of the potential barrier, (say 0.7 volts for silicon and 0.3 volts for germanium), the opposition of the potential barriers will be overcome and current starts to flow.

It occurs because the negative voltage repels or pushes the electrons towards the junction by giving them the energy to cross over and combine with the holes being pushed in the opposite direction approaching the junction by the positive voltage. This results in a characteristics curve of zero current flowing up to this voltage point, known as the knee point on the static curves and then high current flows through the diode with a little increase in the external voltage as shown below.

Forward Characteristics Curve for a Junction Diode

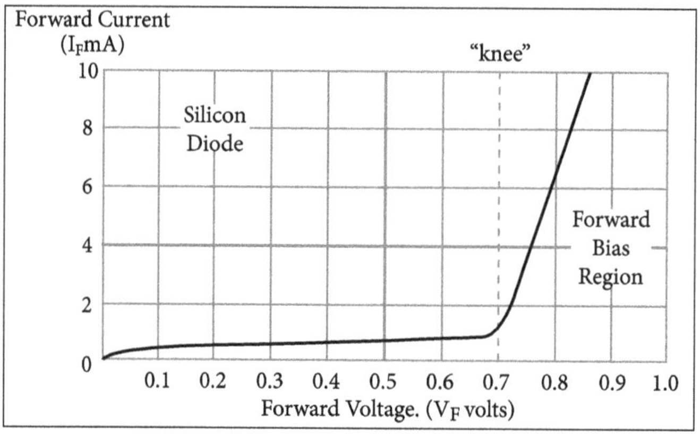

The application of forward biasing voltage on the junction diode will result in the depletion layer becoming very thin and narrow, which represents a low impedance path through the junction thereby allowing high currents to flow. The point at which this sudden increase in current that takes place is known as the knee point and is represented on the static I-V characteristics curve.

5.1.3 Formation of Depletion Zone

Depletion region refers to a region where flow of charge carriers are decreased over a given time and eventually results in empty mobile charge carriers or full of immobile charge carriers.

p-type and n-type Semiconductor

The p-type semiconductor is formed by adding trivalent impurities to pure or intrinsic semiconductor, whereas the n-type semiconductor is formed by adding pentavalent impurities to the pure or intrinsic semiconductor.

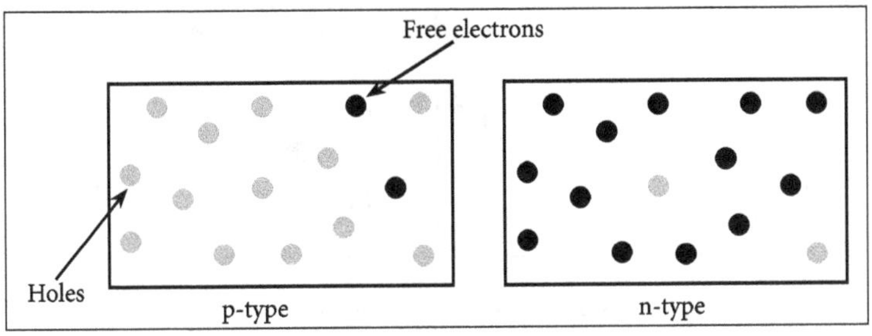

In p-type semiconductor, holes are the majority carriers, whereas free electrons are the minority carriers. On the other hand, in n-type semiconductors, free electrons are the majority carriers, whereas holes are the minority carriers.

Flow of free Electrons and Holes

If p-type semiconductor is joined with n-type semiconductor, then p-n junction is formed. The region in which the p-type and n-type semiconductors are joined is called as p-n junction. This p-n junction separates the n-type semiconductor from p-type semiconductor.

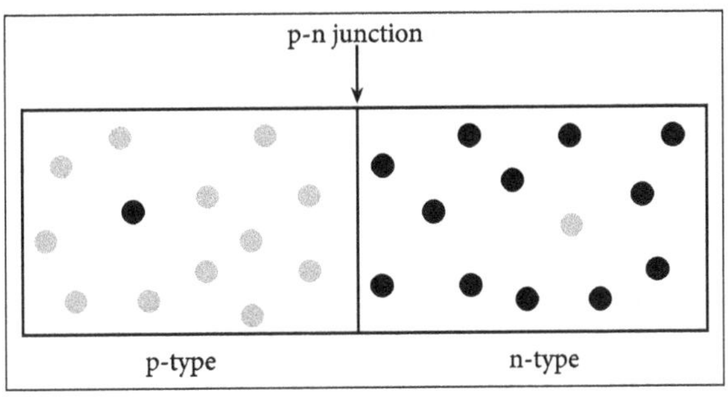

In n-type semiconductor, large numbers of free electrons are present; due to this they get repelled and try to move from a region of high concentration to a region of low concentration. Moreover, near the junction, free electrons and holes are close to each other. According to Coulomb's law, there exist forces of attraction between the opposite charges. Hence, the free electrons from n-side get attracted towards the holes at the p-side. Thus, the free electrons move from n-side to p-side. Similarly, holes move from p-side to n-side.

Positive and Negative Charge at p-n Junction

The free electrons that are crossing the junction from the n-side provide extra electrons to the atoms on the p-side by filling holes in the p-side atoms. The atom that gains extra electron at p-side has more number of electrons than protons. When the atom gains an extra electron from the outside atom it will become a negative ion.

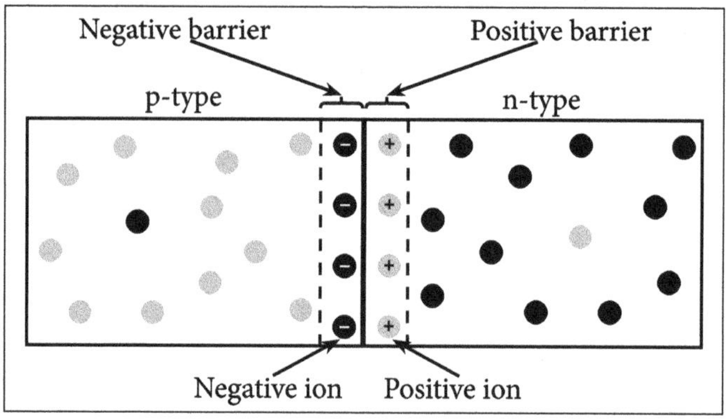

Thus, each free electron that is crossing the junction from n-side to fill the hole in the p-side creates a negative ion at the p-side. Similarly, each free electron that left the parent atom at n-side to fill the hole in the p-side creates a positive ion at n-side.

Negative ion has more number of electrons than protons. Hence, it is negatively charged. Thus, a net negative charge is built at the p-side of the p-n junction. Similarly, positive ion has more number of protons than electrons. Hence, it is positively charged. Thus, a net positive charge is built at the n-side of the p-n junction.

The net negative charge at p-side of the p-n junction prevents the flow of free electrons crossing from n-side to p-side, because the negative charge present at the p-side of the p-n junction repels the free electrons. Similarly, the net positive charge at n-side of the p-n junction prevents the flow of holes from p-side to n-side.

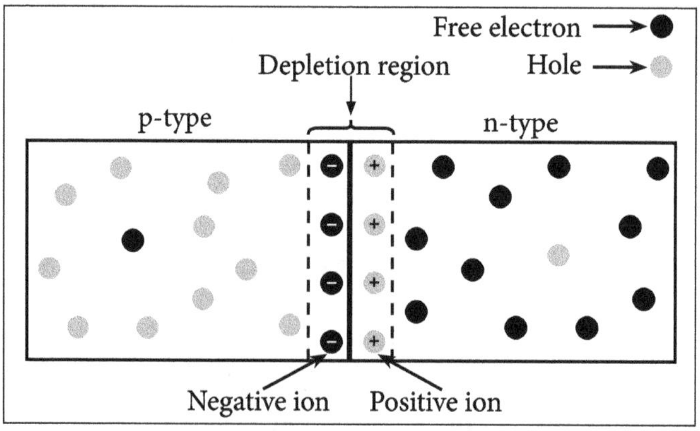

Thus, immobile positive charge at n-side and immobile negative charge at p-side near the junction acts like a wall or barrier and prevents the flow of free electrons and holes. The region near the junction where flow of charge carriers are decreased over a given time and eventually results in empty charge carriers or full of immobile charge carriers is known as depletion region.

The depletion region is also known as depletion layer, depletion zone, space charge region or space charge layer. The depletion region acts like a wall between the p-type and n-type semiconductor and prevents the flow of free electrons and holes.

Knee voltage of diode: The voltage above which the diode operates in normal region is called as Knee voltage of a diode.

5.1.4 V-I Characteristics

V-I Characteristics of a Diode Under Forward bias

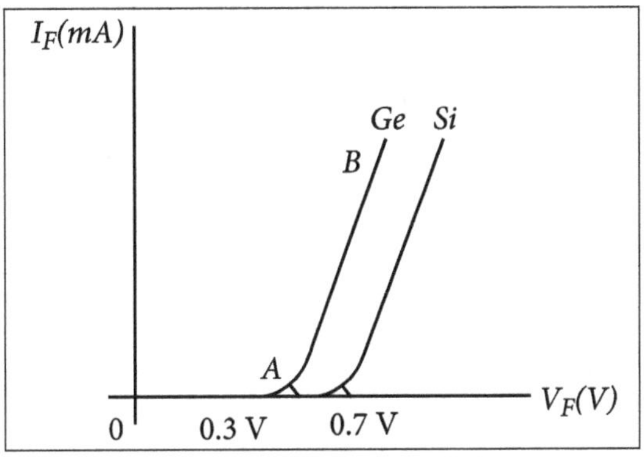

Diode under forward condition.

When the negative terminal of the battery is connected to the n-type and positive terminal of the battery connected to the p-type of p-n junction, the bias applied is known as forward bias.

As the forward biased voltage (V_F) increased for $V_F < V_o$, the forward current (I_F) is almost zero (region OA) because the potential barrier prevents the holes from p-region and electrons from n-region to flow across the depletion region in the opposite direction.

For $V_F > V_o$, the potential barrier at the junction completely disappears. Cut-in or threshold voltage is 0.3V for Germanium and 0.7V for silicon.

V-I Characteristics of a Diode Under Reverse Bias

Reverse bias condition normally generates holes in p-regions which are attracted towards the negative terminals of the battery and the electron in the n-region are attracted towards the positive terminal of the battery.

The minority carriers (i.e electrons in the p-region and holes in the n-region) cross over the junction and flow towards the majority carrier sides giving rise to a small reverse current known as Reverse saturation current I_o.

The free electron from the n-type moving towards the positive terminals of the battery acquires sufficient energy to move with high velocity to dislodge valence electrons from semiconductor atoms in the crystal. The reverse voltage at which the junction breakdown occurs is known as breakdown voltage.

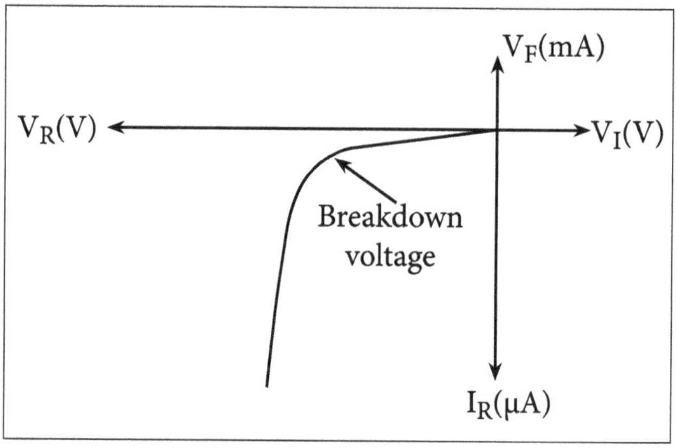

V-I Characteristics under reverse bias.

5.2 Zener Breakdown

The Zener Breakdown is observed in the Zener diodes having V_z less than 5V or between 5 to 8 volts. When a reverse voltage is applied to a Zener diode, it causes a very intense electric field to appear across a narrow depletion region. Such an intense electric field is strong enough to pull some of the valence electrons into the conduction band by breaking their covalent bonds, these electrons then become free electrons which are available for conduction. A large number of such free electrons will constitute a large reverse current through the Zener diode and breakdown is said to have occurred due to the Zener effect.

Breakdown Characteristic of a Zener Diode

Figure shows the reverse characteristic of a Zener diode. It may be noted from above figure that as the reverse voltage (V_R) starts to increase the reverse current (usually called Zener current, I_Z) remains negligibly small up to the Knee of the curve.

At this point, the effect of Breakdown process begins, at this point, the voltage is called Zener Breakdown voltage or simply Zener voltage V_z and it remains essentially constant. This ability of diode is called regulating ability and is important feature of Zener diode. It maintains, essentially a constant voltage across its terminals over a specified range of Zener current values.

Two important point about the characteristics of Zener diode:

- There is a maximum value of Zener current designated as I_{ZM} or $I_{Z(max)}$ above which diode may be damaged. The value of this current is given by the maximum power dissipation of Zener diode. As long as the maximum power dissipation is not exceeded the diode will not be damaged.

- There is a minimum value of Zener current called break over current designated as I_{ZK} or $I_{Z(min)}$ which must be maintained in order to keep the diode in breakdown (or regulation). When the current is reduced below the knee of the curve, the voltage changes drastically and the regulation is lost.

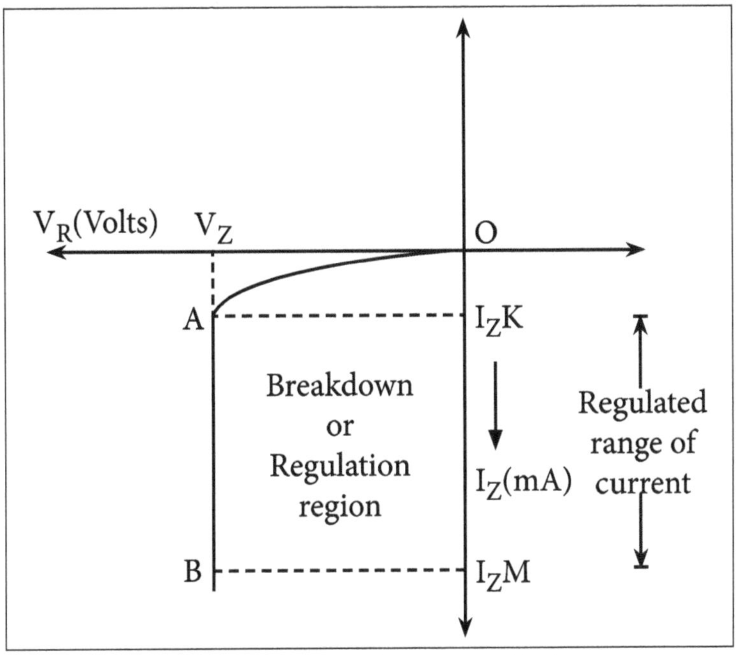

Reverse characteristic of a Zener diode.

5.2.1 Avalanche Breakdown and its Reverse Characteristics

Avalanche Breakdown

When electrons and holes in semiconductors are accelerated by applied voltage, these accelerated holes and electron collide with the bound electrons and produces free electrons. This causes even more collision due to increase in current value. This is called as Avalanche breakdown.

With increasing electric field, electrons travelling through the depletion layer gains higher and higher kinetic energy. Some of them will have sufficient energy to raise an electron from the valence band to the conduction band, thereby creating an electron-hole pair as shown in figure. This phenomenon is called as impact ionization. The electrons and holes created by impact ionization are themselves also accelerated by the electric field.

Consequently, the electron and the holes and the original carrier can create even more carriers by impact ionization. The result is similar to a snow avalanche on a mountainside. When p reaches crit, the carrier creation rate and the reverse current rises abruptly. This is called avalanche breakdown. If the ϕ_{bi} term in equation (11) is ignored,

$$V_B = \frac{\varepsilon_S \xi_{crit}^2}{2qN} \quad \text{...(11)}$$

E_{crit} is about 5×10^5 V/cm at N = 10^{17} cm^{-3} and is approximately proportional to $N^{0.2}$. Therefore equation (12) may be reduced to:

$$V_B = \frac{\varepsilon_S \xi_{crit}^2}{2qN} - \phi_{bi} \quad \text{...(12)}$$

$$V_B(V) \approx 15 \times \left(\frac{10^{17}}{N}\right)^{0.6} \propto 1/N^{0.6}$$

V_B is about 15 V at N = 10^{17} cm^{-3}.

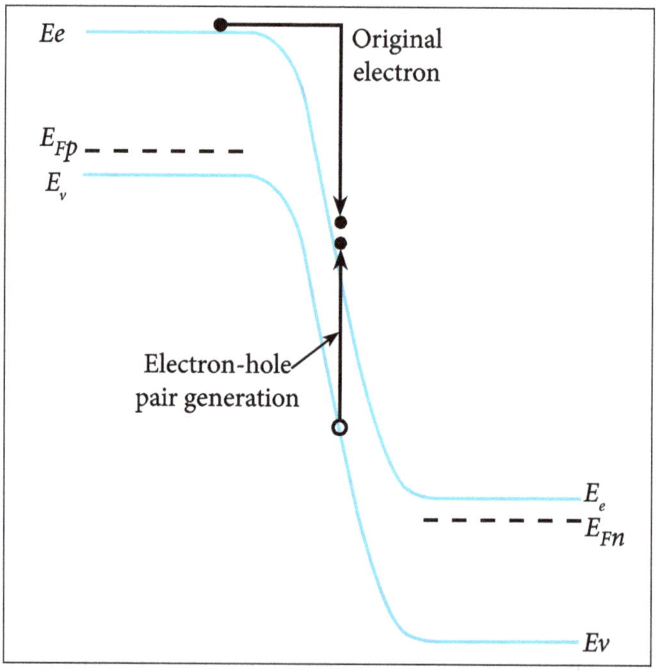

Electron-hole pair generation by impact ionization. The incoming electron gives up its kinetic energy to generate an electron-hole pair.

In general, it is necessary and effective to reduce the junction doping concentration if a larger breakdown voltage is desired. The energy that a carrier must possess in order to initiate impact ionization increases with increasing band gap, E_g. Thus, the breakdown voltage, for given N_a and N_d, progressively increases from Ge to Si to Ga As diodes, due to increasing E_g.

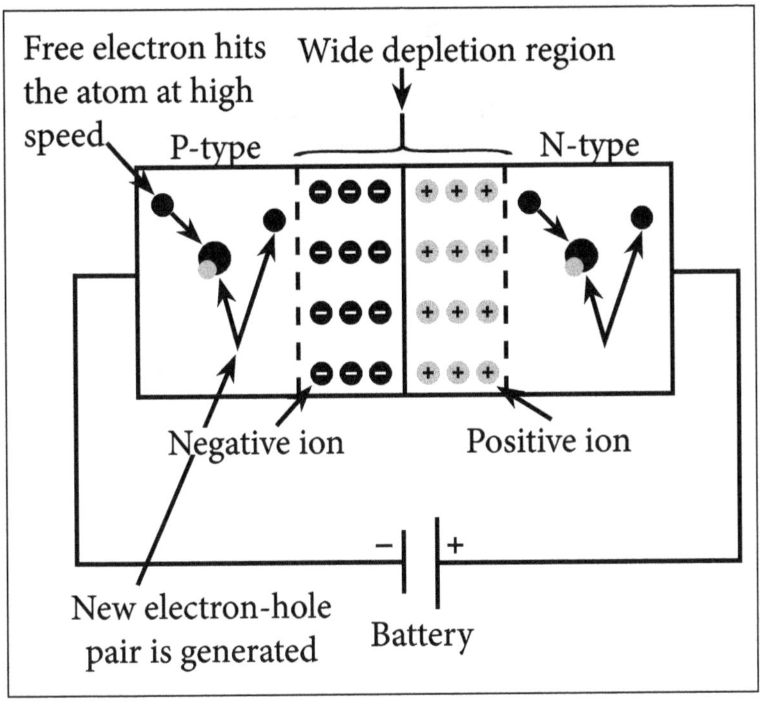

Avalanche breakdown.

Characteristics

The generation of minority carriers in the reverse biased p-n junction diode does not depend on the applied reverse voltage. It depends only on the temperature. However, the applied reverse voltage on the p-n junction diode supplies energy to the minority carriers. The free electrons, which gain energy from the external voltage, are accelerated to greater velocities. If the voltage applied to the p-n junction diode is further increased to a higher value, the free electrons gain large amount of energy and travel at very high speed.

When these high speed free electrons collide with the ions, atoms or valence electrons, they transfer their kinetic energy to the valence electrons. The valence electron, which gains enough energy from the high speed free electron will break the bonding with the parent atom and becomes a free electron. The point at which the free electron is left is called as hole. Thus, a free electron at the conduction band and a hole at the valence band is generated as pair.

The newly generated free electron gains energy from the external voltage source and travel at high speed. If this newly generated free electron collides with an ion or atom, another free electron and hole is generated. Likewise, a large number of minority carriers are generated with few free electrons. This process is called carrier multiplication. In other words, carrier multiplication is the process by which collision of single free electron with the ion or atom leads to the generation of multiple free electrons and holes.

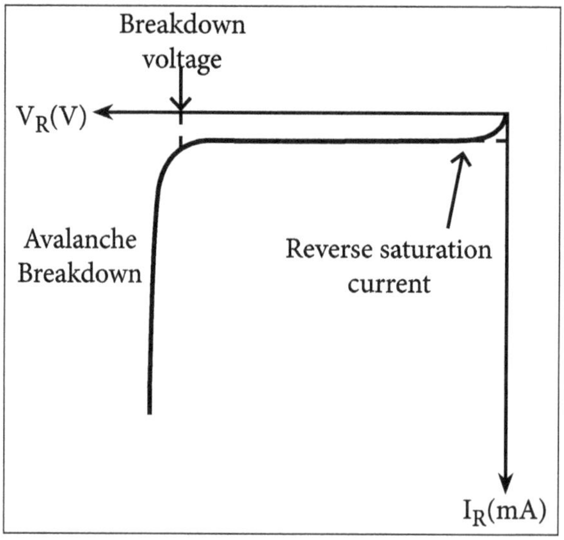

Reverse characteristics.

Thus, at high reverse voltage, large number of minority carriers are generated. This large number of minority carriers cause sudden increase in reverse current, which leads to junction breakdown. In other words, the sudden rise in reverse current destroys the depletion region.

5.2.2 Junction Capacitance

Depletion Layer or Transition Capacitance, C_T

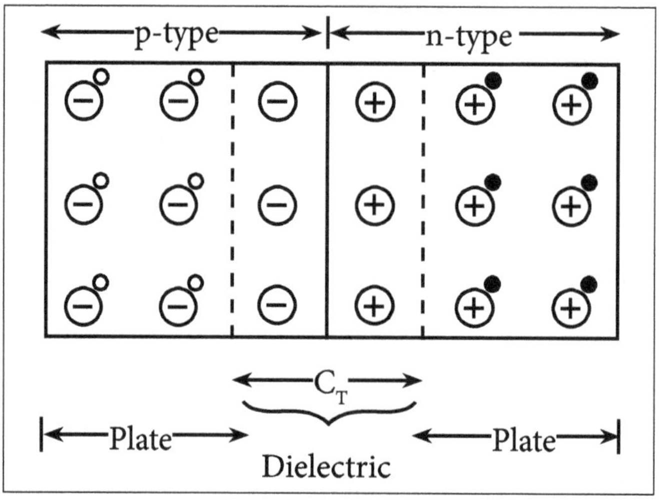

Depletion layer capacitance.

When a p-n junction is formed, a layer of positive and negative ions called depletion layer is formed on either side of the junction. Since this capacitance is formed in the junction area, therefore, it is also called space charge capacitance. This capacitance arises due to immobile charges at the junction varying with the applied voltage, so that this is simply called as junction capacitance. Figure above shows the formation of depletion layer capacitance.

The capacitance of a parallel plate capacitor is given by the equation $C = \frac{\varepsilon_0 . A}{d}$ where ε_0 is the permittivity of dielectric (insulator) between the plates of area A separated by a distance d. Since the depletion layer width d increases with the increase in reverse bias voltage, the resulting depletion layer capacitance will decrease with the increased reverse bias.

The depletion layer capacitance depends upon the nature of a p-n junction, semiconductor material and magnitude of the applied reverse voltage. It is given by the relation:

$$C_T = \frac{C_o}{\left(1+\frac{V_r}{V_T}\right)^n} \quad ...(13)$$

Where,

C_o = Capacitance at zero bias condition.

V_T = Volt equivalent of temperature

$$n = \begin{cases} \frac{1}{2} \text{ for step or abrupt alloy junction} \\ \frac{1}{3} \text{ for linearly graded junction or diffused junction} \end{cases}$$

It is evident from the above relation that the value of depletion layer capacitance (C_T) can be controlled by varying the applied reverse voltage. This property of variable capacitance, possessed by a reverse biased PN-junction is used in the construction of a device called varicap of varactor, which is chiefly used in FM circuits. Range of depletion layer capacitance is approximately in nF (nanofarad).

5.2.3 Varactor Diode

The varactor diode is a semiconductor, voltage dependent variable capacitor alternatively known as varicap or voltacap or Voltage Variable Capacitor (VVC) or tuning diode.

Basically, it is just a reverse biased junction diode, whose mode of operation depends on its transition capacitance (C_T). A varactor diode is specially constructed to have high resistance under reverse bias. Capacitance for varactor diode is in the range of Picofarad.

In a conventional diode, the depletion region exists between p-region and n-region as shown in the below figure.

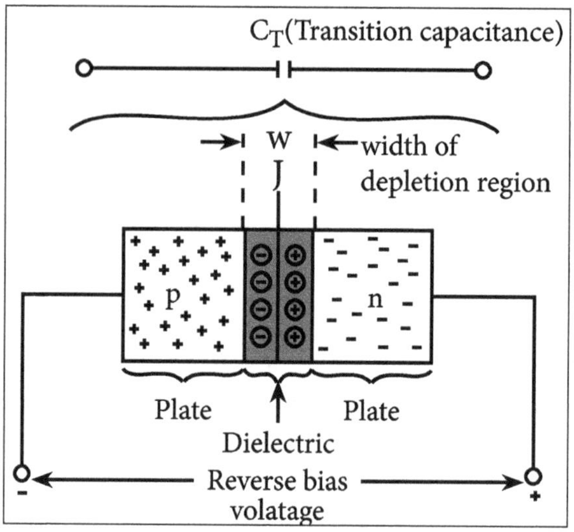

p-n junction reverse biased.

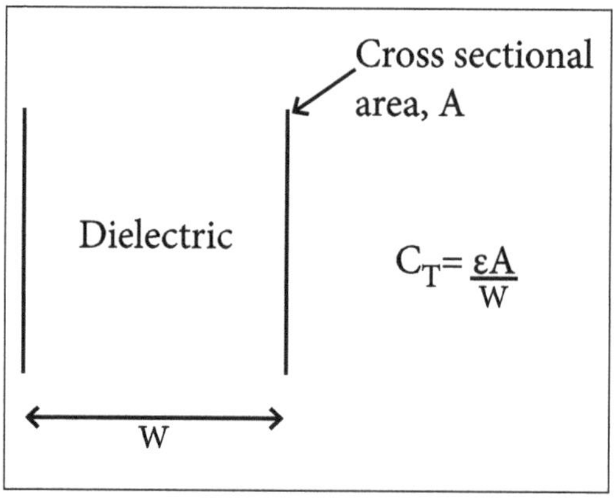

p-n junction-effective parallel plate capacitor.

Holes are the majority carriers in p-type semiconductor material and electrons are the majority charge carriers in n-type semiconductor material. There is a neutral space charge region or the depletion between these two which do not have any majority carriers. There are two plates in the capacitor between which there is a dielectric material.

The p-region and n-region act like the plates of capacitor, while the depletion region acts like a dielectric. Hence, there exists a capacitance at the p-n junction called transition capacitance C_T. It is denoted by the expression:

$$C_T = \frac{\varepsilon A}{W}$$

Where,

C_T = Total capacitance of the junction.

C = Permittivity of the semiconductor material.

A = Cross-sectional area of the junction.

W = Width of the depletion layer.

In the below figure the depletion layer is formed of a p-n junction. This depletion layer can also be made of a MOS or a Schottky diode. This is very important in MMIC and CMOS technology.

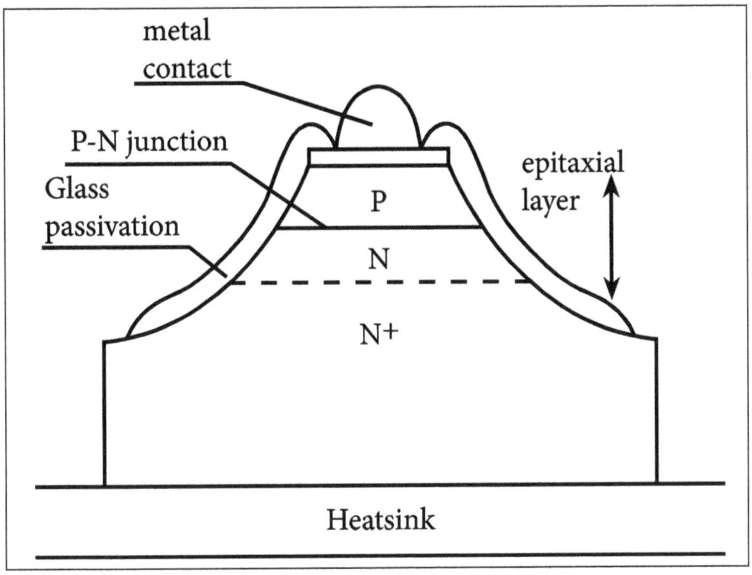

Internal structure of a varactor diode.

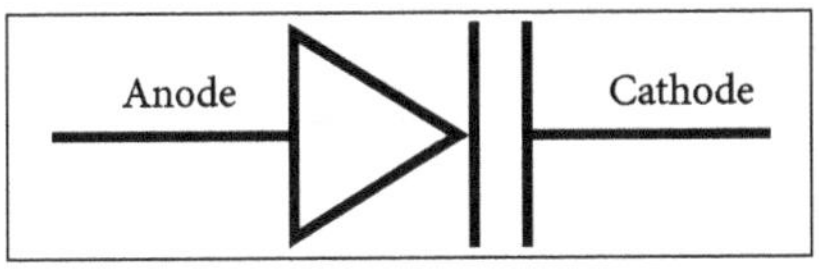

Symbol.

Operation of Varactor Diode

Varactors are operated in the reverse biased state. No current flows, but since the thickness of the depletion region varies with the applied bias voltage, the capacitance of the diode can be made to vary.

Generally, the depletion region thickness is directly proportional to the square root of the applied voltage. Its capacitance is inversely proportional to the depletion region thickness. Thus, the capacitance is inversely proportional to the square root of the applied voltage.

All diodes exhibit this phenomenon to some degree, but varactor diodes are manufactured specifically to exploit this effect and increase the capacitance. The capacitance can be controlled by the applied voltage. Curve between reverse bias voltage V_R across the varactor diode and total junction capacitance C_T is shown in the below figure.

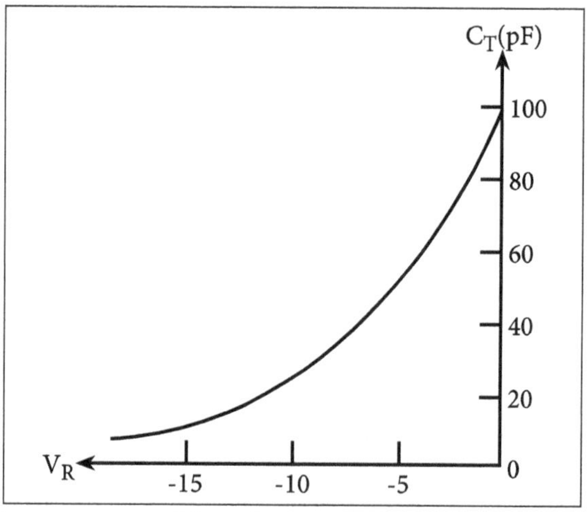

Curve between reverse bias voltage V_R and junction capacitance C_T.

Varactors may be of two types. The doping profile of:

- The abrupt junction diode.
- The hyper abrupt junction diode.

The abrupt junction diode has uniform doping and a capacitive tuning ratio (TR) of 4:1. The hyper abrupt junction diode has highest impurity concentration near the junction. It results in narrow depletion layer and larger capacitance. Also, changes in V_R produce larger changes in capacitance. Such a diode has a tuning range of 10:1, which is enough to tune a broadcast receiver through its frequency range of nearly 3:1.

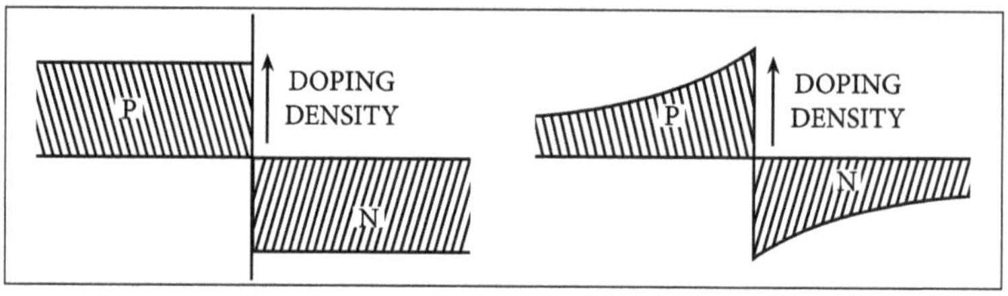

Doping profile of abrupt junction and hyper abrupt junction.

Applications of varactor diode:

- Tuned circuits.
- FM modulators.
- Television receivers.
- Parametric amplifiers.
- Automatic frequency control devices.
- Adjustable band pass filters.

5.3 Simple Diode Circuits and Load Line

A diode is used as a part of an electronic circuit and the diode current i_D depends on other circuit elements. A simple diode circuit is shown in the below figure. Applying Kirchhoff's voltage law, we can express the source voltage V_S and the diode current i_D by:

$$V_S = v_D + R_L i_D$$

Which gives the diode current i_D as:

$$i_D = \frac{V_S - v_D}{R_L} \quad ...(14)$$

Since, the diode is forward biased, the diode current i_D is related to the diode voltage v_D by the Shockley diode equation:

$$i_D = I_S\left(e^{v_D/nV_T} - 1\right) \quad ...(15)$$

Which shows that i_D depends on v_D, which in turn depends on i_D.

Thus, equations (14) and (15) can be solved for v_D and i_D by any of the following methods:

- Graphical method.
- Approximate method.
- Iterative method.

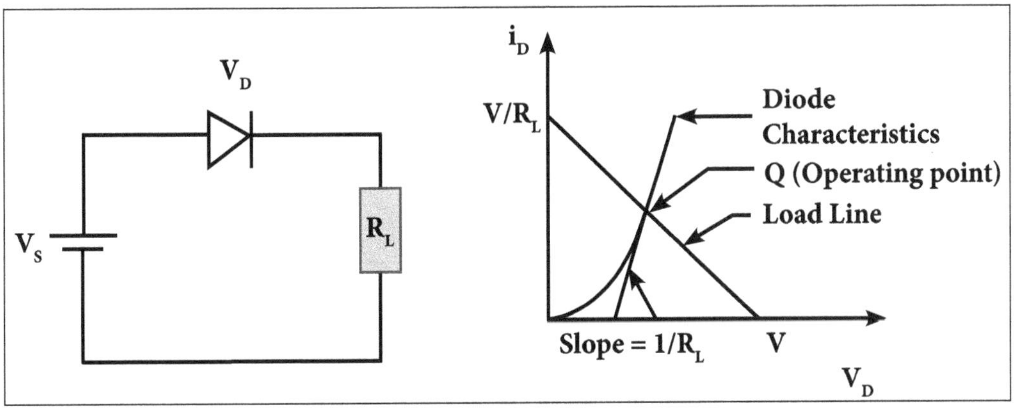

The line of intersection of Q points is called as DC load line. The straight line represented by the above graph is known as the load line. The load line passes through two points, $I = 0$, $v_D = V_S$ and $v_D = 0$, $I = V_S/R_L$.

The intersection point of load line and diode characteristics curve gives the operating point.

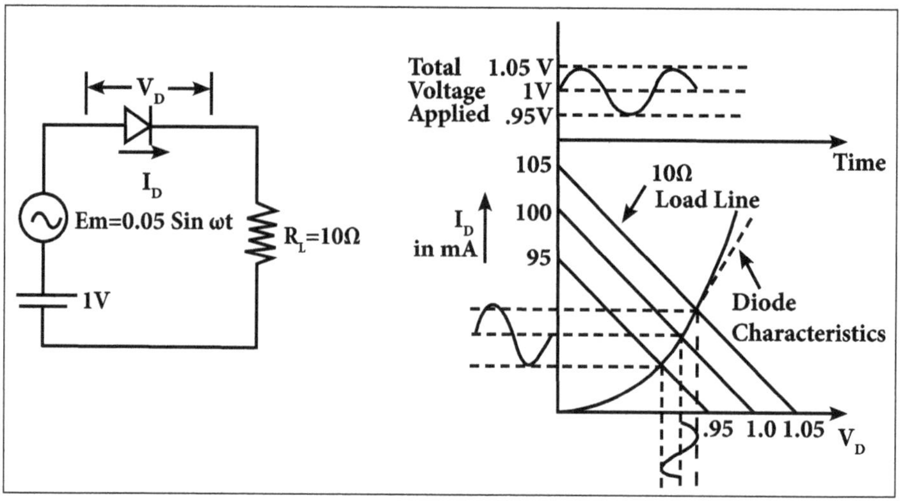

5.3.1 Linear Piecewise Model

In figure the exponential curve is approximated by two straight lines, line A with zero slope and line B with a slope of $1/r_D$. It can be seen that for the particular case shown in figure, over the current range of 0.1 mA to 10 mA the voltages predicted by the straight-lines model shown differ from those predicted by the exponential model by less than 50 mV.

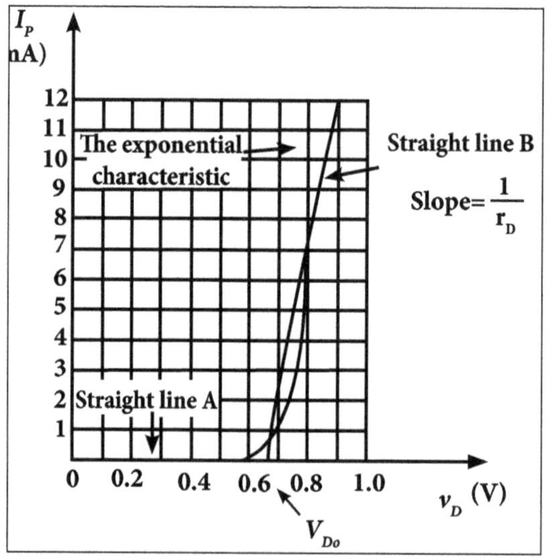

Approximating the diode forward characteristics with two straight lines: piecewise linear model.

The choice of these two straight lines is not unique, one can obtain a closer approximation by restricting the current range over which the approximation is required. The straight-lines or piecewise-linear model of the above figure can be given by:

$$i_D = 0, v_D \leq v_{DO}$$

$$i_D = \frac{v_D - v_{Do}}{r_D}, v_D \geq v_{Do}$$

Where,

V_{Do} - The intercept of line B on the voltage axis.

r_D - The inverse of the slope of line B.

5.4 Rectifier and its Properties

Rectifier is defined as an electronic device used for converting ac voltage into unidirectional voltage. A rectifier is a unidirectional conduction device like a p-n junction diode. The two types of rectifiers are:

Half-Wave Rectifier

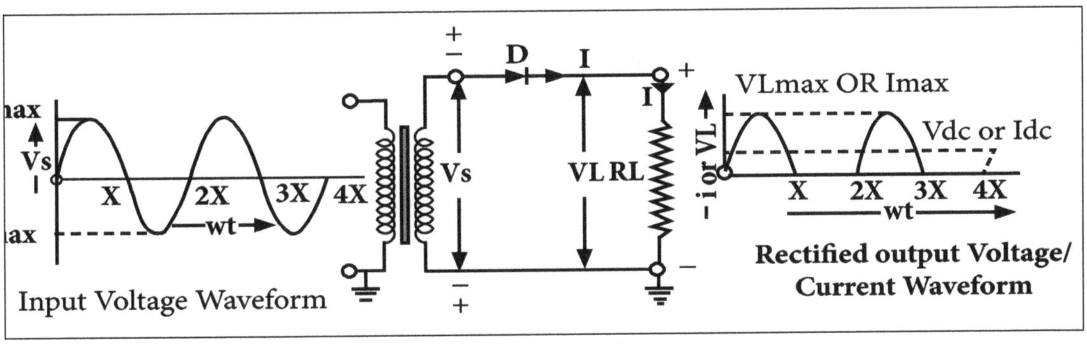

Half Wave Rectifier.

The primary of the transformer is connected to ac supply. This induces an ac voltage across the secondary of the transformer. During the positive half cycle of the input voltage, the polarity of the voltage across the secondary forward biases the diode. As a result, a current I_L flows through the load resistor, R_L.

The forward biased diode offers a very low resistance and hence the voltage drop across it is very small. Thus the voltage appearing across the load is practically the same as the input voltage at every instant.

At the negative half cycle of the input voltage, the polarity of the secondary voltage gets reversed. As a result, the diode is reverse biased.

Experimentally no current flows through the circuit and almost no voltage is developed across the resistor. All the input voltage appears across the diode itself. Hence, we conclude that when the input voltage is going through its positive half cycle, the output voltage is almost the same as the input voltage and during the negative half cycle no voltage is available across the load.

It explains the unidirectional pulsating dc waveform obtained as output. The process of removing one half of the input signal to establish a dc level is aptly called half wave rectification.

If the input voltage reaches its maximum value V_m during the negative half cycle, the voltage across the diode is also maximum. Hence, this maximum voltage is known as the peak inverse voltage.

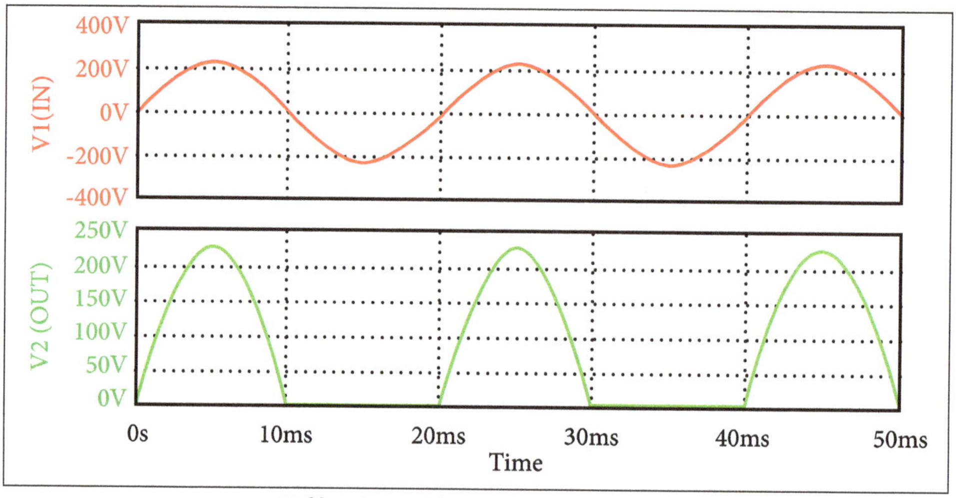

Half wave rectifier - output waveform.

The average of load current (I_{dc}):

$$I_{dc} = \frac{1}{2\pi}\int_0^{\pi} I_m \sin\omega t = \frac{I_m}{\pi}$$

Let, the load current be $i_L = I_m \sin\omega t$.

Ripple Factor of Half Wave Rectifier

Ripple factor is defined as the ratio of rms value of ac component to the dc component in the output.

$$\text{Ripple factor}(r) = \frac{\left(I_{rms}^2 - I_{dc}^2\right)^{\frac{1}{2}}}{I_{dc}} = 1.21$$

The rms value of the load current (I_{rms}),

$$I_{rms} = \frac{I_m}{4}$$

Rectification efficiency is defined as the ratio between the output power to the ac input power.

Efficiency, η

DC power delivered to the load/AC input power from the transformer = P_{dc}/P_{ac}.

DC power delivered to the load, $P_{dc} = I_{dc}^2 R_L = (I_{max}/pi)^2 R_L$.

AC power input to the transformer,

P_{ac} = Power dissipated in diode junction + Power dissipated in load resistance $R_L = I_{rms}^2 R_F + I_{rms}^2 R_L = \{I_{max}^2/4\}[R_F + R_L]$.

So, Rectification Efficiency, $\eta = P_{dc}/P_{ac} = \{4/2\}[R_L/(R_F + R_L)] = 0.406/\{1 + R_F/R_L\}$

The maximum efficiency that can be obtained by the half wave rectifier is 40.6%. This is obtained if R_F is neglected.

Full Wave Rectifier

A Full wave rectifier is a circuit which allows a unidirectional current flow through the load during the entire input cycle. There are two types such as center-tapped and bridge rectifier.

Bridge Rectifier

The need for center-tapped power transformer is eliminated in the bridge rectifier.

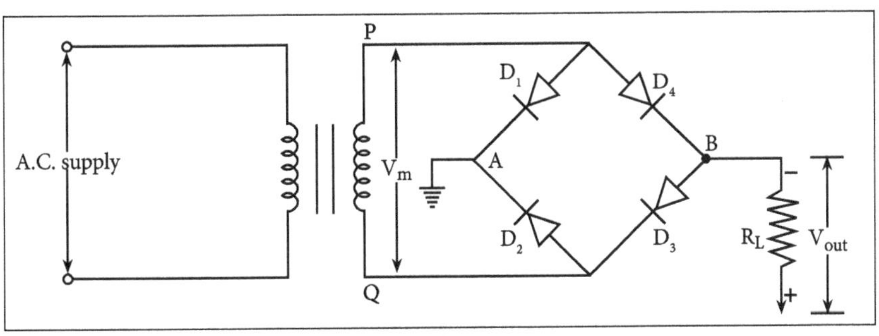

Circuit diagram.

It contains four diode D_1, D_2, D_3 and D_4 connected to form bridge. The a.c supply to be rectified is applied to diagonally opposite ends of the bridge through the transformer between the load R_L connected.

During the positive half-cycle P–positive and Q–negative, D_1 and D_3 are in forward bias, D_2 and D_4 are in reverse bias.

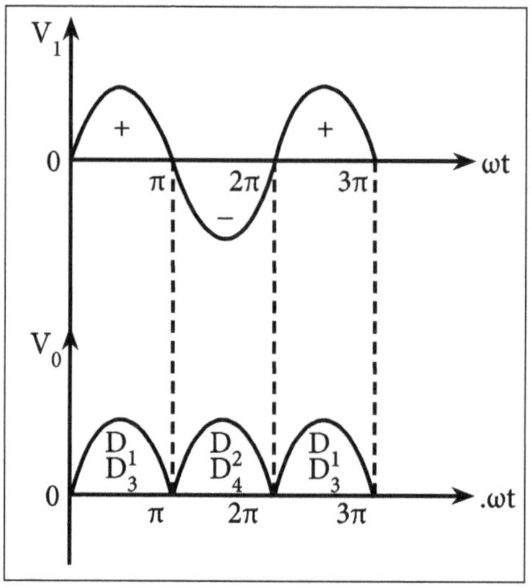

Input-Output Waveforms.

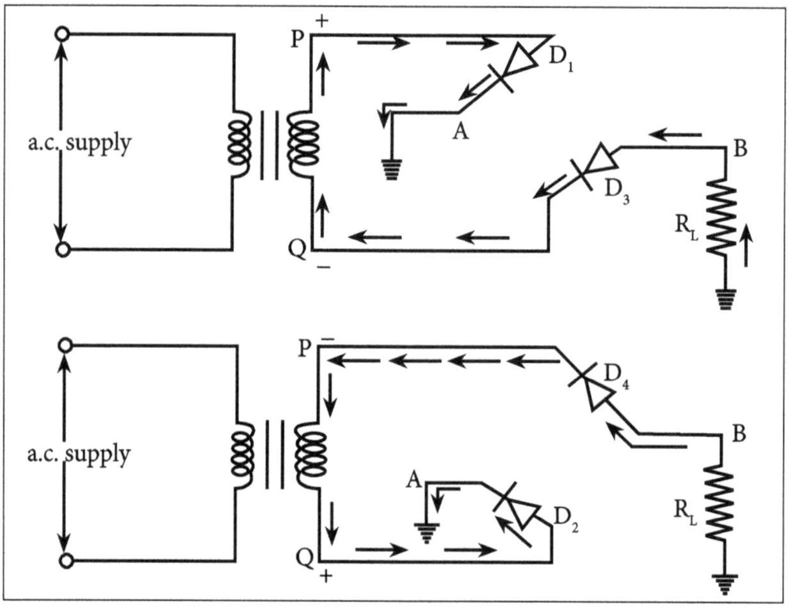

Bridge rectifier in positive and negative half-cycle condition.

During the Negative half-cycle P–negative and Q–positive, D_1 and D_3 are in reverse bias and D_2 and D_4 are in forward bias conditions. These two diodes will be in series through the load R_L, the current flows from A to B through the load in the same directions, d.c output is obtained across load R_L.

Transformer Utilization Factor (TUF)

$$TUF = \frac{\text{d.c power delivered to the load}}{\text{a.c rating of the transformer}}$$

$$= \frac{P_{dc}}{P_{ac}\text{rated}}$$

$$= \frac{\left(\frac{2I_m}{\pi}\right)^2 .R_L.\left(\frac{2I_m}{\pi}\right)^2 .R_L}{\frac{V_m}{\sqrt{2}}.\frac{I_m}{\sqrt{2}} \quad \frac{I_m^2}{2}.R_L} \left[\because V_m = I_m R_L\right]$$

$$TUF = \frac{8}{\pi^2}$$

$$TUF = 0.81 \text{ or } 81\%$$

Peak Inverse Voltage (PIV)

It is defined as the maximum reverse voltage that a diode can withstand without destroying the junction. As calculated, PIV = $2V_m$. The same procedure is repeated when D_1 if OFF and D_2 is ON.

Center-Tapped Full Wave Rectifier

In full wave rectification, current flow through the load in the same direction for both half-cycles of input a.c voltage. It uses center-tapped transformer which provide equal voltage above and below the center-tapped for both the half-cycles.

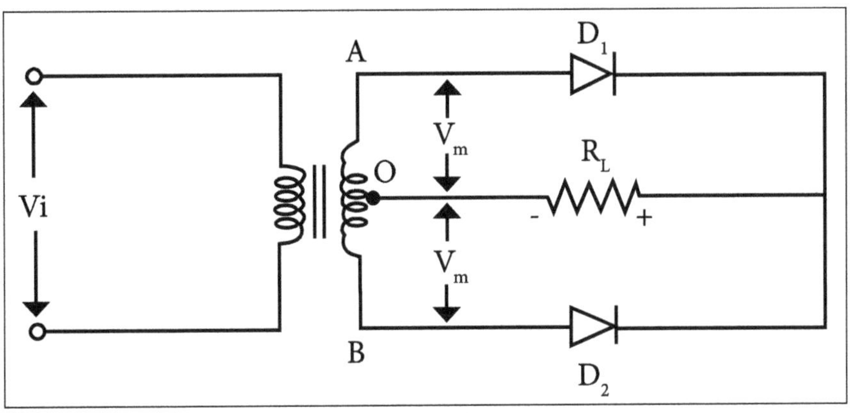

Center-tapped FWR.

The center-tap on the secondary winding of a transformer is taken as the ground or zero voltage reference point. The circuit uses two diodes, which are connected to the center-tapped secondary winding of the transformer. Diode D_1 utilizes the ac voltage appearing across the upper half (OA) of secondary winding for rectification while diode D_2 uses the lower half winding (OB).

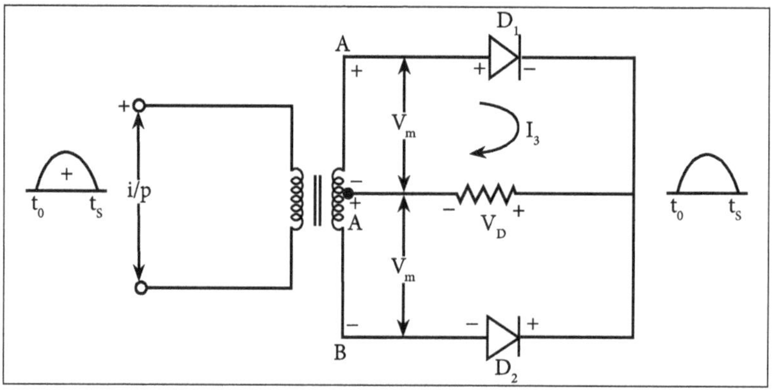

During the half cycle.

During the positive half-cycle of secondary voltage, the end A of the secondary winding becomes positive and end B negative as shown in above figure. This makes the diode D_1 forward biased and D_2 reverse biased. Therefore diode D_1 conducts, while diode D_2 does not. The conventional current flow through diode D_1, load resistor R_L and upper half of secondary winding.

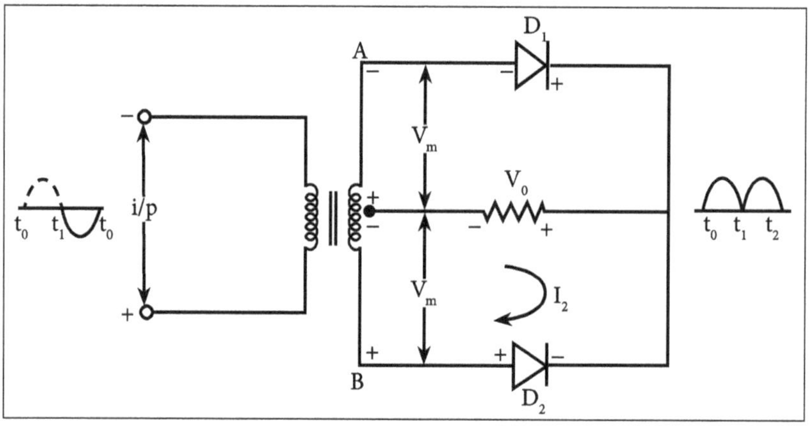

During the -ve half cycle.

During the Negative half-cycle, end A of the secondary winding becomes negative and end B positive as showed in figure. Therefore, diode D_2 conduct while diode D_1 does not. The conventional current flow is through diode D_2, load R_L and lower half winding. The current in the load R_L is in the same direction for both half-cycle of input a.c. voltage.

Average and DC Value

$$I_{dc} = \frac{\text{Area under the curve over a half cycle}}{\text{Base or period}}$$

$$= \frac{1}{\pi}\int_0^{\pi} I_L . d\omega t$$

$$I_{dc} = \frac{1}{\pi}\int_{0}^{\pi} I_m \sin\omega t.d\omega t$$

$$= \frac{I_m}{\pi}\left[-\cos\omega t\right]_0^{\pi}$$

$$= \frac{I_m}{\pi}\left[\cos\pi - \cos\theta\right]$$

$$I_{dc} = \frac{2I_m}{\pi}$$

Similarly,

$$V_{dc} = \frac{2V_m}{\pi}$$

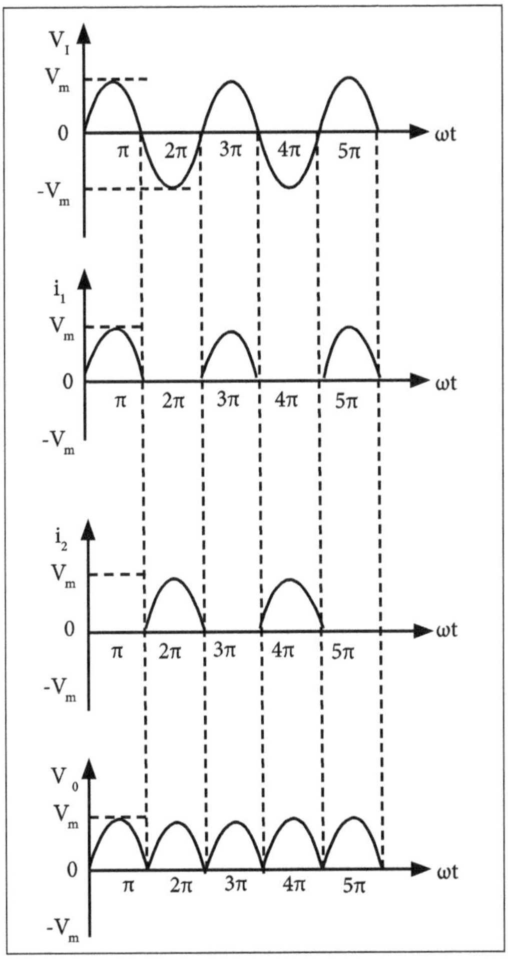

Input and output Wave forms.

RMS Value of Load Current

$$I_{rms} = \sqrt{\frac{1}{\pi}\int_0^{\pi}(I_m \sin\omega t)^2 .d\omega t}$$

$$= \sqrt{\frac{I_m^2}{\pi}\int_0^{\pi}\left(\frac{1-\cos 2\omega t}{2}\right)d\omega t}$$

$$= \sqrt{\frac{I_m^2}{2\not{\pi}}.\not{\pi}}$$

$$I_{rms} = \frac{I_m}{\sqrt{2}}$$

$$V_{rms} = \frac{V_m}{\sqrt{2}}$$

Ripple Factor

$$\gamma = \frac{V_{r.rms}}{V_{dc}} = \frac{\sqrt{V_{rms}^2 - V_{dc}^2}}{V_{dc}}$$

$$= \sqrt{\left(\frac{V_m}{\sqrt{2}}.\frac{\pi}{2V_m}\right)^2 - 1}$$

$$\gamma = 0.483 \text{ or } 48.3\%$$

Efficiency

$$\eta = \frac{P_{dc}}{p_{ac}} = \frac{I_{dc}^2.R_L}{I_{rms}^2\left(R_L + R_s + R_f\right)}$$

$$I_{dc} = \frac{2I_m}{\pi} \text{ and } I_{rms} = \frac{I_m}{\sqrt{2}}$$

$$\eta = \frac{\left(\frac{2I_m}{\pi}\right)^2 .R_L}{\left(\frac{I_m}{\sqrt{2}}\right)^2\left(R_L + R_s + R_f\right)}$$

$$= \frac{8}{\pi^2} \cdot \frac{R_L}{(R_L + R_s + R_f)}$$

If $(R_s + R_f) << R_L$

Then $\eta = 0.812$ or 81.2%

Transformer Utilization Factor (TUF)

$$(TUF)s = \frac{P_{dc}}{P_{dc} \cdot \text{rated}}$$

$$= \frac{I_{dc}^2 R_L}{V_{rms} \cdot I_{rms}} = \frac{(2I_m / \pi)^2 . R_L}{\frac{V_m}{\sqrt{2}} \cdot \frac{I_m}{\sqrt{2}}}$$

$$= \frac{8}{\pi^2} = 0.811 \quad (\text{or}) \quad 81.1\%$$

Advantages:

- The output voltage and transformer efficiency are high.
- Lower ripple factor.
- Higher TUF.

Disadvantages:

- Usage of additional diode and bulky transformer is needed and hence increase in cost.
- PIV of diode is high (i.e., $2V_m$).
- The output voltage is half of the secondary voltage.

5.4.1 Idea of Regulation

The variation of dc output voltage as a function of dc load current is called regulation. The variation of V_{dc} with I_{dc} for a half wave rectifier is obtained as follows:

$$I_{dc} = \frac{I_m}{\pi} = \frac{V_m / \pi}{R_f + R_L}$$

But, $V_{dc} = I_{dc} \times R_L$

$$V_{dc} = \frac{V_m}{\pi}\left[\frac{R_L}{R_f + R_L}\right]$$

$$= \frac{V_m}{\pi}\left[1 - \frac{R_f}{R_f + R_L}\right]$$

$$= \frac{V_m}{\pi} - I_{dc}R_f$$

$$\therefore V_{dc} = \frac{V_m}{\pi} - I_{dc}R_f$$

This result shows that V_{dc} equals V_m/π at no load and that the dc voltage decreases linearly with an increase in dc output current. The larger the magnitude of the diode forward resistance, the greater is this decrease for a given current change.

6

Bipolar Junction Transistors

6.1 Formation of PNP/NPN Junctions

A bipolar junction transistor is a three terminal semiconductor device consisting of two p-n junctions that is able to amplify or "magnify" a signal. It is a current controlled device.

The three terminals of the BJT are the collector, the base and the emitter. A signal of small amplitude if applied to the base is available in the amplified form at collector of the transistor. This is the amplification provided by the BJT.

The three parts of a BJT are collector, emitter and base. Initially, before knowing about the bipolar junction transistor characteristics, we must know about the modes of operation for this type of transistors.

The modes are:

- Common Base (CB) mode.
- Common Emitter (CE) mode.
- Common Collector (CC) mode.

Bipolar transistor has three regions: Emitter, Base and collector.

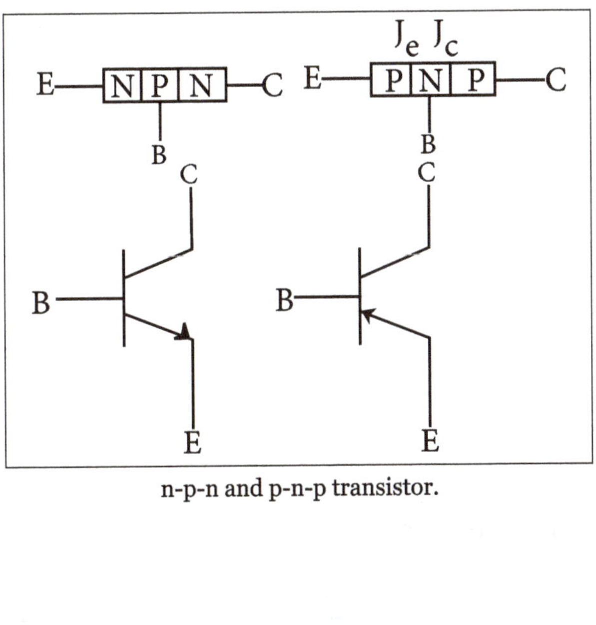

n-p-n and p-n-p transistor.

N-P-N Bipolar Junction Transistor

In n-p-n bipolar transistor, one p-type semiconductor resides between the two n-type semiconductors. The diagram below is a n-p-n transistor. Now I_E, I_C is the emitter current and collector current respectively and V_{EB} and V_{CB} are emitter base voltage and collector base voltage respectively.

According to the convention, if the emitter, base and collector current I_E, I_B and I_C current goes into the transistor, sign of the current is taken as positive and if the current goes out from the transistor, then sign is taken as negative.

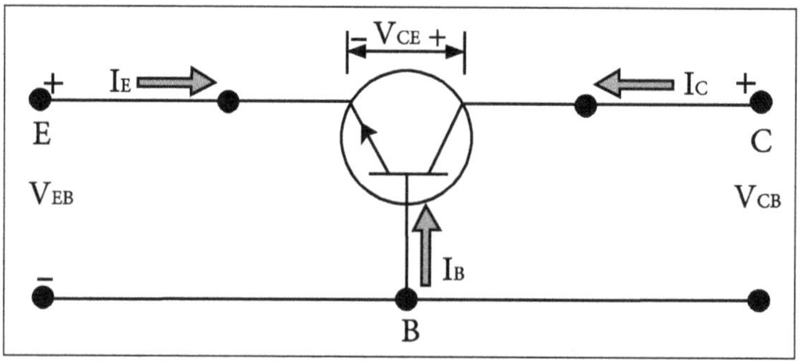

NPN transistor.

We could tabulate the different currents and voltages inside the n-p-n transistor.

Transistor type	I_E	I_B	I_C	V_{EB}	V_{CB}	V_{CE}
n - p - n	-	+	+	-	+	+

P-N-P Bipolar Junction Transistor

Similarly for p-n-p bipolar junction transistor, the n-type semi-conductors is sandwiched between two p-type semiconductors. The diagram of a p-n-p transistor is as shown below.

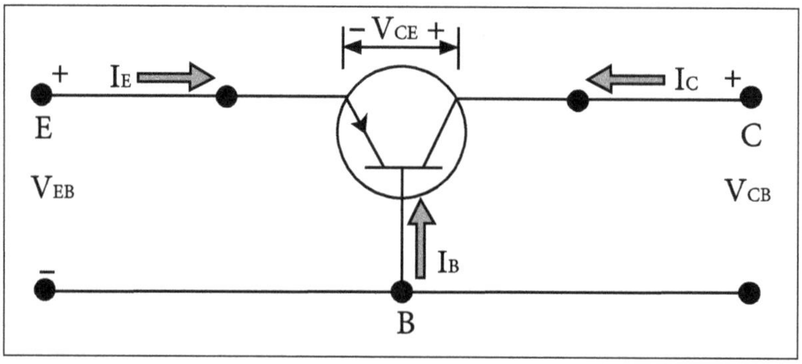

PNP transistor.

For p-n-p transistors, current enters into the transistor through the emitter terminal. Like any BJT, the emitter-base junction is forward biased and the collector-base junction is reverse biased. We could tabulate the emitter, base and collector current as well as the emitter base, collector base and the collector emitter voltage for p-n-p transistors.

Transistor type	I_E	I_B	I_C	V_{EB}	V_{CB}	V_{CE}
p - n - p	+	-	-	+	-	-

Energy band Diagram

N-P-N Transistor

The energy band diagram of an n-n-n transistor is shown in the below figure. At the instant of formation of the junctions, diffusion of majority carriers occur and the energy bands in each region get shifted. The energy bands in n-region move down while those in p-region are shifted upward. The displacement of energy bands and diffusion of majority charge carriers come to a halt as soon as the Fermi levels in the three regions are equalized.

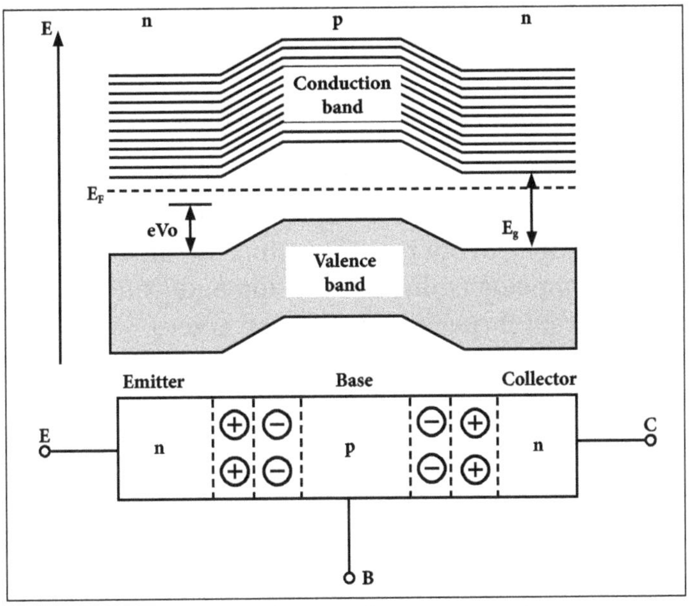

Energy Band Diagram of an n-p-n transistor.

The process of relative displacement of energy bands causes bending of bands in the junction regions and creation of potential barriers of height eVo. The potential barriers inhibit the majority carrier diffusion but promote minority carrier drifting. In equilibrium condition, the current components due to diffusion and drift of carriers balance each other. Hence, the net current through each junction is zero.

P-N-P Transistor

The energy band diagram of an p-n-p transistor is shown in the below figure. At the instant of formation of junctions, diffusion of majority carriers occurs and the energy bands in each region get shifted. The energy bands in n-region move down while those in p-region are shifted upward. The displacement of energy bands and diffusion of majority charge carriers come to a halt as soon as the Fermi levels in the three regions are equalized.

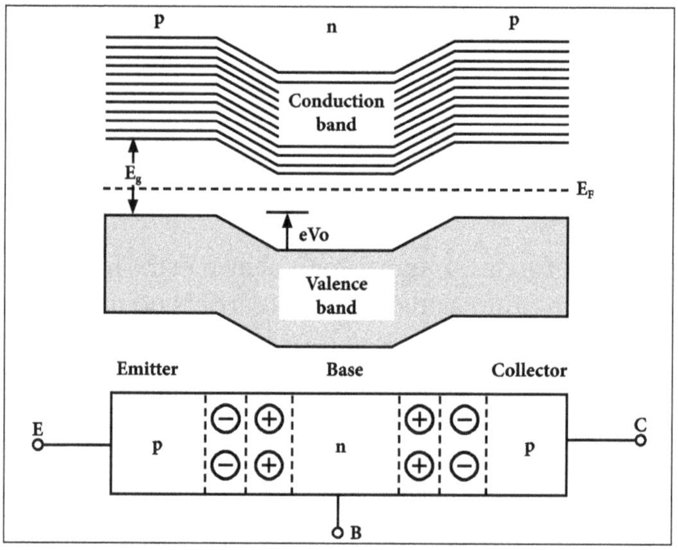

Energy band diagram of a p-n-p transistor.

The process of relative displacement of energy bands causes bending of bands in the junction regions and creation of potential barriers of height eV_o. The potential barriers inhibit majority carrier diffusion, but promote minority carrier drilling. In equilibrium condition, the current components due to diffusion and drift of carriers balance each other. Hence, the net current through each junction is zero.

Transistor Mechanism

The basic operation of the transistor is described using the p-n-p transistor. The p-n junction of the transistor is forward-biased, whereas the base-to-collector is without a bias, as shown in the below figure.

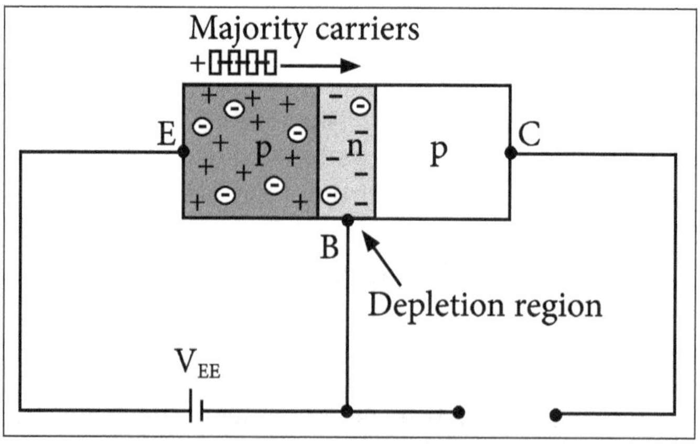

Forward-biased junction of a p-n-p transistor.

The middle portion is termed as the base (B), while the two end portions are known as the emitter (E) and collector (C). The junction between the emitter and the base is called as the emitter-base junction or the emitter junction (J_E). The junction between the collector and the base is called as the collector-base junction or the collector junction (JC).

The depletion region gets reduced in width due to the applied bias, resulting in a heavy flow of majority carriers from p-type to the n-type material gushing down the depletion region and reaching the base. The forward bias on the emitter-base junction will cause current to flow. This flow of current consists of two components:

- Holes injected from emitter to base.
- Electrons injected from base to emitter.

The design of the transistor ensures that the base region is fabricated very lightly compared to the emitter or collector regions. Consequently, we have the holes injected from the emitter into the base in large numbers and we neglect the injection of electrons from the base region. Let us now remove the base-to-emitter bias of the p-n-p transistor, as shown in the below figure.

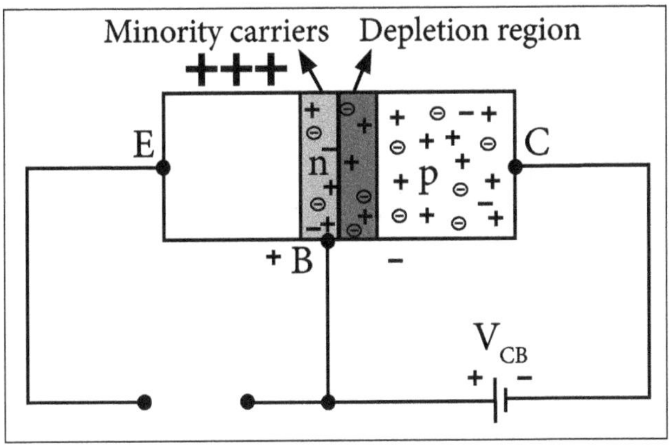

Reverse-biased junction of a p-n-p transistor.

The flow of majority carriers is zero, resulting in a minority-carrier flow. Thus, one p-n junction of a transistor is reverse-biased, while the other is kept open. The operation of this device becomes much easier when they are considered as separate blocks. The drift currents due to thermally generated minority carriers have been neglected, since they are very small.

6.1.1 CE, CB and CC Configuration

Transistor Configurations

As the bipolar transistor is a three terminal device, there are basically three possible ways to connect it within an electronic circuit with one terminal being common to both the input and output. Each method of connection responds differently to its input signal within the circuit, as the static characteristics of the transistor vary with each circuit arrangement.

- Common base configuration has voltage gain but it has no current gain.
- Common emitter configuration has both current and voltage gain.
- Common collector configuration has current gain but it has no voltage gain.

Common Base (CB) Configuration

In the common base or grounded base configuration, the base connection is common to both the input signal and the output signal, with the input signal being applied between the base and the emitter terminals. The corresponding output signal is taken between the base and the collector terminals as shown, with the base terminal grounded or connected to a fixed reference voltage point.

The input current flowing into the emitter is large, since it is a sum of both the base current and the collector current respectively. Hence, the collector current output is less than the emitter current input resulting in the current gain for this type of circuit of 1 or less. In other words, common base configuration attenuates the input signal.

Common Base Transistor Circuit

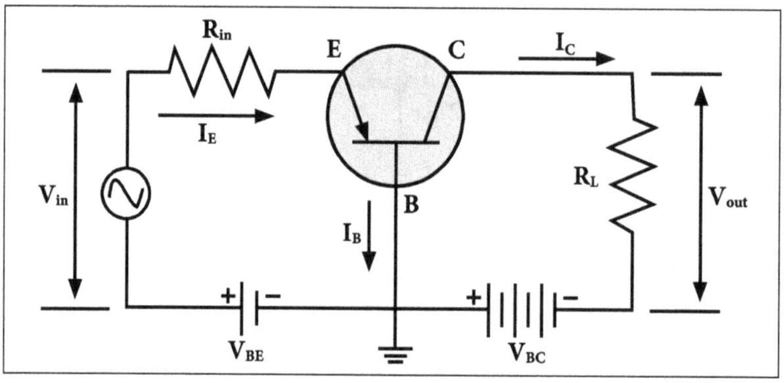

CB transistor.

In this type of non-inverting amplifier configuration, the signal voltages V_{in} and V_{out} are in-phase. This is not a common transistor arrangement, as it posses unusually high voltage gain characteristics. Its output characteristics represent that of a forward biased diode and the input characteristics represent that of an illuminated photodiode.

Common Base Voltage Gain

This type of bipolar transistor configuration has high ratio of output to input resistance or particularly load resistance (R_L) to input resistance (R_{in}) giving it a value of resistance gain. Then the voltage gain (A_v) for the common base configuration is given as:

$$A_V = \frac{V_{out}}{V_{in}} = \frac{I_C \times R_L}{I_E \times R_{in}}$$

Where,

I_C / I_E = Current gain, (α)

R_L / R_{in} = Resistance gain

The common base circuit is usually used only in single stage amplifier circuits like, microphone pre-amplifier or radio frequency amplifiers due to its very good high frequency response.

Common Emitter (CE) Configuration

In the common emitter or grounded emitter configuration, the input signal is applied to the base, while the output is taken from the collector and emitter as shown in the below figure. This is the most commonly used circuit configuration for transistor based amplifiers. It represents the normal method of bipolar transistor connection.

The common emitter amplifier configuration produces the highest current and power gain of all the three bipolar transistor configurations. This is mainly because the input impedance is low as it is connected to the forward biased PN junction, whereas the output impedance is high as it is taken from a reverse biased PN junction.

Common Emitter Amplifier Circuit

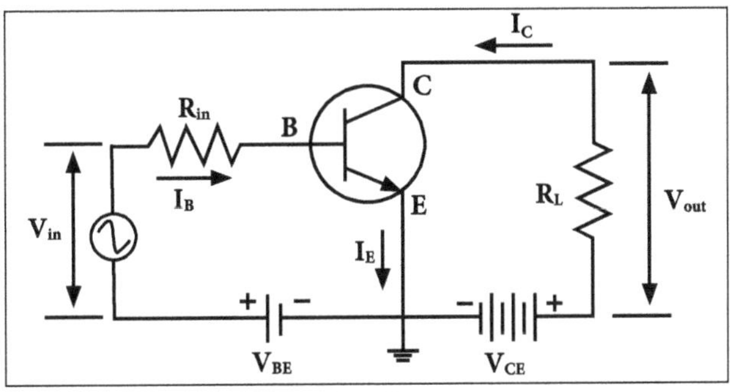

CE transistor.

In this type of configuration, the current flowing out of the transistor should be equal to the current flowing into the transistor as the emitter current is given by, $I_E = I_C + I_B$.

As the load resistance (R_L) is connected in series with the collector, the current gain of the common emitter transistor configuration is quite large, it is a ratio of I_C/I_B and it is denoted by the Greek symbol β. As the emitter current for the common emitter configuration is defined as, $I_E = I_C + I_B$, the ratio of I_C/I_E is denoted by the Greek symbol α.

Since the electrical relationship between these three currents, I_B, I_C and Ie is determined by the physical construction of the transistor itself, any small change in the base current (I_B) will result in much larger change in the collector current (I_C).

Thus, small changes in the current flowing in the base will control the current in the emitter-collector circuit. Typically, β has a value between 20 and 200 for most general purpose transistors. So, if the transistor has a β value of 100, then one electron will flow from the base terminal for every hundred electrons flowing between the emitter-collector terminals.

By combining the expressions for α and β, the mathematical relationship between these parameters and hence the current gain of the transistor can be written as:

$$\alpha = I_C / I_E \text{ and } \beta = I_C / I_B$$

$$\therefore I_C = \alpha I_E = \beta I_B$$

From current equations we know that:

$$\alpha = \beta / (\beta + 1) \beta = \alpha / (1 - \alpha)$$

$$I_E = I_C + I_B$$

Where,

I_C = Current flowing into the collector terminal.

I_B = Current flowing into the base terminal.

I_E = Current flowing out of the emitter terminal.

This type of bipolar transistor configuration has larger input impedance, current and power gain than that of the common base configuration while its voltage gain is much lower. The common emitter configuration is an inverting amplifier circuit, which means that the resulting output signal is 180° out of phase with the input voltage signal.

Common Collector (CC) Configuration

The common collector or grounded collector configuration is one in which the collector is common through the supply. The input signal is connected directly to the base, while the output is taken from the emitter load as shown in the below figure. This type of configuration is commonly known as the voltage follower or emitter follower circuit.

The common collector or emitter follower configuration is very useful for impedance matching applications, because of its very high input impedance in the region of hundreds of thousands of Ohms while having relatively low output impedance.

Common Collector Transistor Circuit

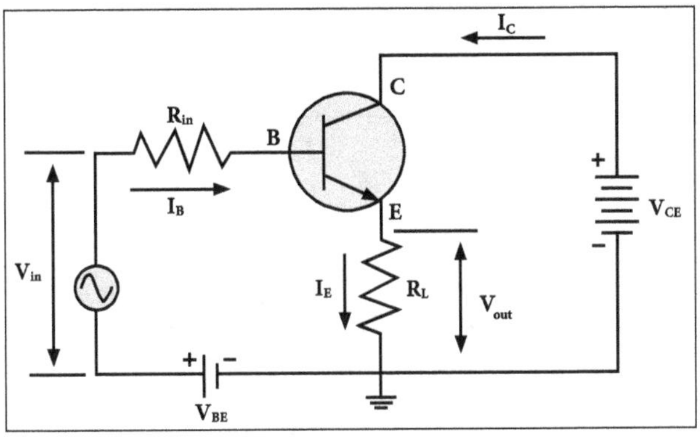

CC transistor.

The common emitter configuration has a current gain approximately equal to the β value of the transistor itself. In the common collector configuration, the load resistance is connected in series with the emitter. So, its current is equal to that of the emitter current.

As the emitter current is the combination of the collector and base current, the load resistance in this type of transistor configuration also has both collector current and input current of the base flowing through it.

Common collector current gain:

$$I_E = I_C + I_B$$
$$A_i = I_E / I_B = (I_C + I_B) / I_B$$
$$A_i = (I_C / I_B) + 1$$
$$A_i = \beta + 1$$

This is a type of non-inverting circuit configuration of bipolar transistor, where signal voltages of V_{in} and V_{out} are in-phase. It has a voltage gain that is always less than 1. The load resistance of the common collector transistor receives both base and collector currents giving a large current gain. Therefore, providing good current amplification with very little voltage gain.

Relationship between DC currents and gains:

$I_E = I_B + I_C$ $I_C = I_E - I_B$ $I_B = I_E - I_C$	$\alpha = I_C / I_E = \beta/(1+\beta)$ $\beta = I_C / I_B = \alpha/(1-\alpha)$

<table>
<tr><td colspan="2">$I_B = I_C/\beta = I_E/(1+\beta) = I_E(1-\alpha)$</td></tr>
<tr><td>$I_C = \beta I_B = \alpha I_E$</td><td>$I_E = I_C/\alpha = I_B(1+\beta)$</td></tr>
</table>

Characteristics of a Transistor

CE Configuration

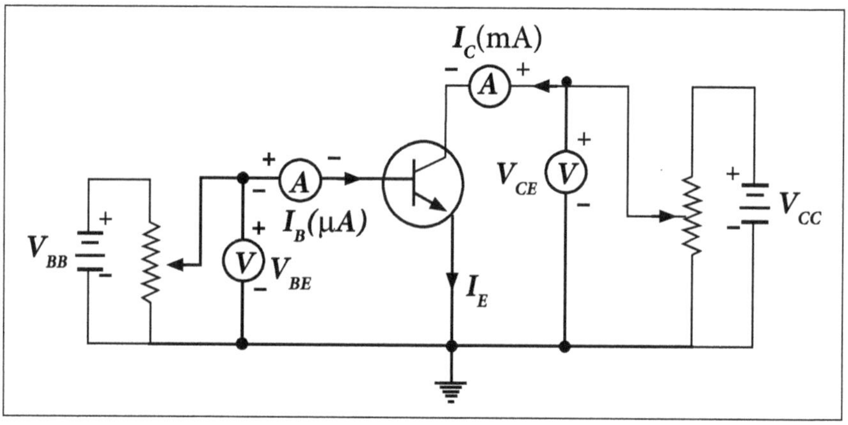

Circuit for CE static characteristics.

Input Characteristics

These characteristics might be obtained by using the circuit arrangement shown in the above figure. First, we adjust the collector to emitter voltage (V_{CE}) to 1 volt. Then, we can increase the base to emitter voltage (V_{BE}) in small suitable steps and record the corresponding values of base current (I_B) at each step.

If we plot a graph with base to emitter voltage (V_{BE}) along the horizontal axis and base current (I_B) along the vertical axis, we shall obtain a curve as shown in the figure. Similar procedure may be used to obtain the characteristics at different values of collector to emitter voltage. The input characteristic gives us the information about the following important point.

- There exists a threshold or knee voltage below which the base current is very small. The value of knee voltage is 0.5 V for silicon and 0.1 V for germanium transistors.
- Beyond the knee point, the base current (I_B) increases with the increase in base to emitter voltage (V_{BE}) for a constant collector to emitter voltage (V_{CE}). However, it may be noted that the value of base current does not increase rapidly as that of the input characteristics of the common base transistor. It means that, input resistance of a transistor in common emitter configuration is higher as compared to the common base configuration.

- As the collector to emitter voltage (V_{CE}) is increased, the curves shift downwards. It occurs because of the fact that, as V_{CE} is increased the depletion width in the base region increases. This reduces the effective base width which in turn reduces the base current.

- The input characteristics may be used to determine the value of a.c input resistance (R_i) of a common emitter transistor. Its value is given by the ratio of change in base to emitter voltage (ΔV_{BE}) to the resulting change in base current (I_B) at a constant collector to emitter voltage (V_{CE}).

Mathematically the a.c input resistance is given by:

$$R_i = (\Delta V_{BE})/(\Delta I_B)$$

The value of a.c input resistance ranges from 600 Ω to 4100 Ω.

It may be noted that, the input characteristic is not linear in the lower region of the curve. Therefore, the input resistance varies with the location of the operating point.

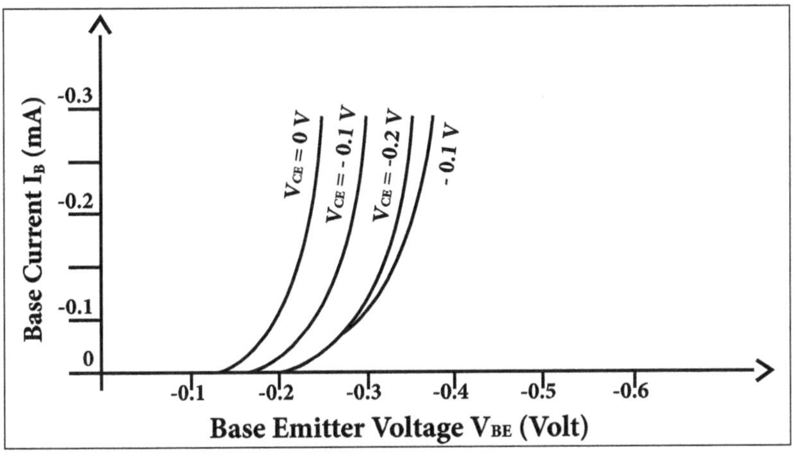

Input characteristics curve.

These characteristics may be obtained by using the circuit shown above. To begin with, increase the base current (I_B) to 40 mA value. Then increase the collector to emitter voltage (V_{CE}) in a number of steps and record the corresponding values of collector current (I_c) at each step.

Output Characteristics

If we plot a graph with collector to emitter voltage (V_{CE}) along the horizontal axis and the collector current (I_c) along the vertical axis, we shall obtain a curve as shown in the figure. A similar procedure may be used to obtain the characteristics at I_B = 80 mA, 120 mA, 160 mA and so on. The output characteristic gives us the information about the following important points.

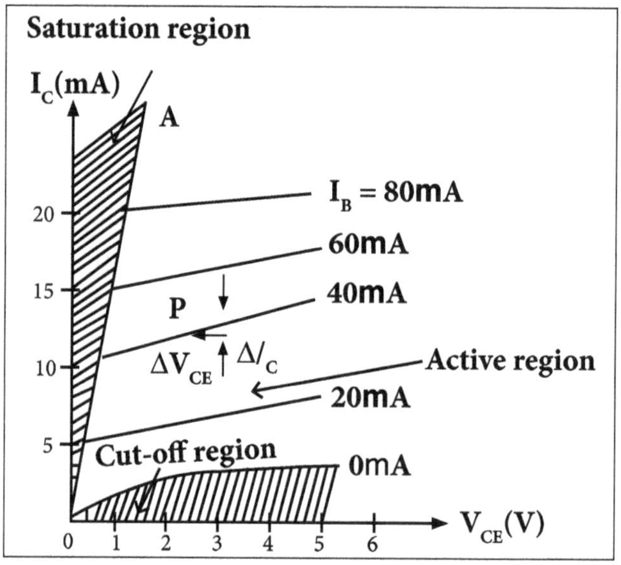

Output characteristic curve.

The output characteristics have three regions namely:

1. Saturation Region

In this region, both the collector and emitter junctions are forward biased and an increase in the base current does not cause a corresponding large change in I_c. The ratio of $V_{CE(sat)}$ to Ic in this region is called as the saturation resistance.

2. Cut-off Region

In this region, both the junctions are reverse biased. Transistor is in OFF state. Hence, the collector current almost becomes zero and the collector voltage almost equals V_{CC}. The transistor is virtually an open circuit between the collector and emitter.

3. Active Region

The central region where the curves are uniform in spacing is called as the active region. It is the region between the saturation and cut-off. In this region, emitter base junction is forward biased and the collector base junction is reverse biased. If the transistor is to be used as a linear amplifier, it should be operated in the active region.

- As the collector to emitter voltage (V_{CE}) is increased above zero, the collector current (I_c) increases rapidly to a saturation value depending upon the value of base current. It may be noted that the collector current (I_c) reaches to a saturation value when V_{CE} is about 1 V.
- When collector to emitter voltage (V_{CE}) is increased further, the collector current slightly increases. This increase in collector current is due to the fact that,

increased value of collector to emitter voltage (V_{CE}) reduces the base current and hence the collector current increases. This phenomenon is called as early effect.

- When the base current is zero, a small collector current exists. This is called as leakage current. However for all practical purposes, the collector current (I_c) is zero when the base current (I_B) is zero, under this condition the transistor is said to be in cut-off.
- The characteristic may be used to determine the value of a.c output resistance (R_0) of a common emitter transistor. Its value at any given operating point (Q) is given by the ratio of change in collector to emitter voltage (ΔV_{CE}) to the resulting change in collector current (ΔI_C) for a constant base current (I_B).

Mathematically, the AC output resistance is expressed as:

$$R_o = \frac{\Delta V_{CE}}{\Delta I_C}$$

The common emitter output resistance of a transistor ranges from 10 kΩ to 50 kΩ.

- The characteristic may be used to determine the small signal common emitter current gain or AC beta (β_0) of a transistor. Thus, if $\Delta I_C = 40 - 20 = 20$ mA $= 20 \times 10^{-3}$ A and $\Delta I_B = 160 - 80 = 80$ μA $= 80 \times 10^{-6}$ A, then small signal common emitter current gain can be written as,

$$\beta_o = \frac{\Delta I_C}{\Delta I_B} = \frac{20 \times 10^{-3}}{80 \times 10^{-6}} = 250$$

CB Configuration

Input Characteristics

To determine the input characteristics, the collector to emitter voltage is kept constant at zero volt and the base current is increased from zero in equal steps by increasing V_{BE}.

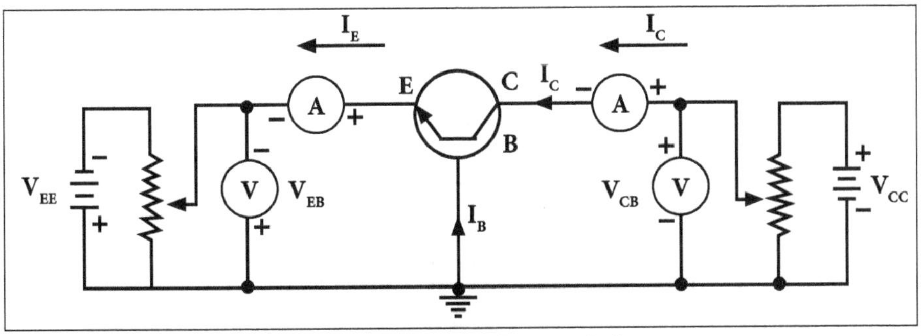

Circuit for CB static characteristics.

The value of V_{BE} is noted for each setting of I_B. This procedure is repeated for higher fixed values of V_{CE} and the curves of I_B vs V_{BE} are drawn.

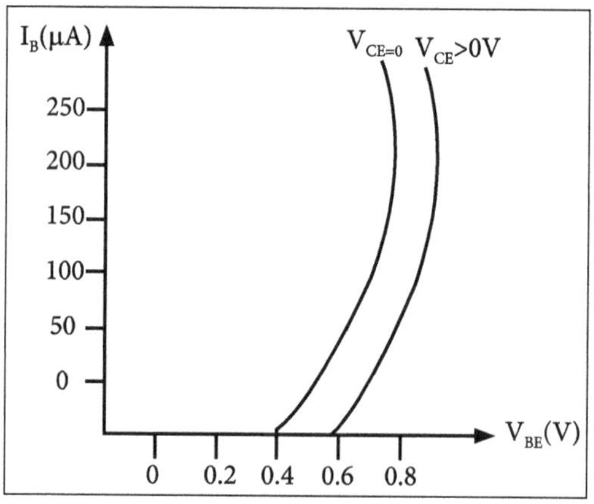

Input characteristic curve.

When V_{CE} = 0, the emitter base junction is forward biased and the junction behaves as a forward biased diode. Hence, the input characteristic for V_{CE} = 0 is similar to that of a forward biased diode.

When V_{CE} is increased, the width of the depletion region at the reverse biased collector base junction will increase. Hence, the effective width of the base would decrease. This effect causes a decrease in the base current I_B. Hence to get the same value of I_B as that for V_{CE} = 0, VBE should be increased. Therefore, the curve shifts to the right as V_{CE} increases.

Output Characteristics

To determine the output characteristics, the emitter current I_E is kept constant at a suitable value by adjusting V_{BE}. The magnitude of V_{CB} is increased in suitable equal steps from zero and the value of I_c is noted for each value of I_E. Now the curves of I_c vs V_{CB} are plotted for different constant values of I_E.

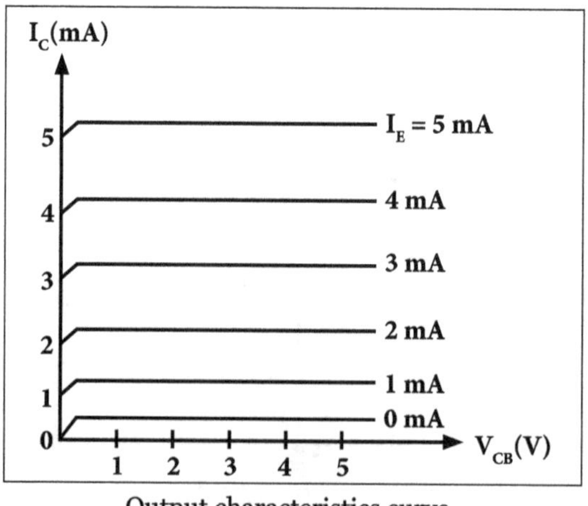

Output characteristics curve.

CC Configuration

Common collector configuration circuit is shown in the below figure. Here the collector is grounded and it is used as a common terminal for both the input and output. It is also called as grounded collector configuration. Base is used as the input terminal, whereas emitter is the output terminal.

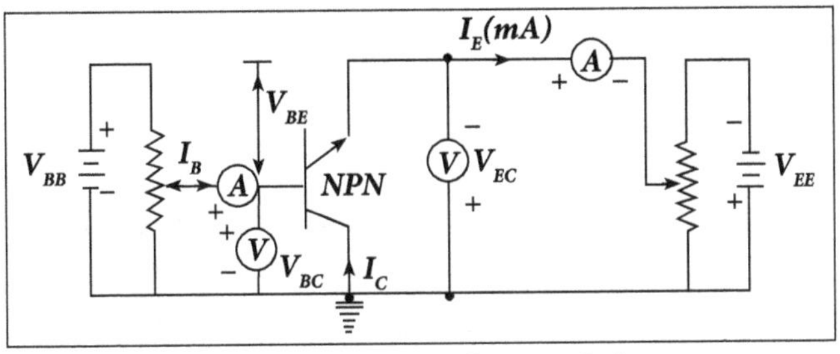

Circuit for CC static characteristics.

Input Characteristics

It is defined as the characteristic curve drawn between input voltages to input current, whereas output voltage is constant.

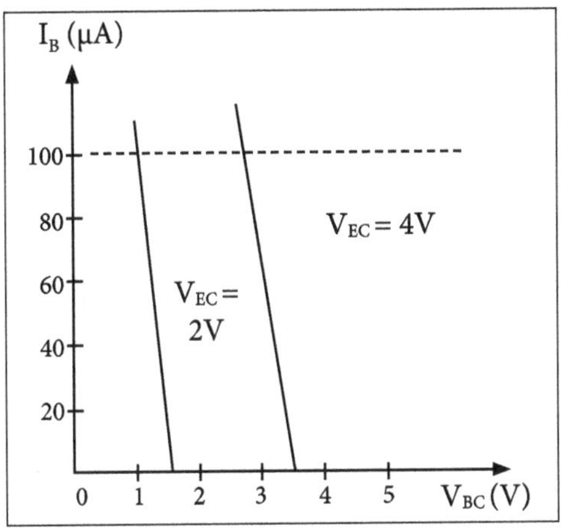

Input characteristic curve.

To determine input characteristics, the emitter base voltage V_{EC} is kept at a suitable fixed value. The base collector voltage V_{BC} is increased in equal steps and the corresponding increase in I_B is noted. This is repeated for higher fixed values of V_{CE}.

A curve is drawn between base current and the base collector voltage at constant collector emitter voltage as shown in the above figure.

Output Characteristics

It is defined as the characteristic curve drawn between output voltages to output current, whereas input current is constant.

To determine the output characteristics, the base current I_B is kept constant at zero and emitter current I_E is increased from zero by increasing V_{EC}. This is repeated for higher fixed values of I_B.

From the characteristic it is seen that, for a constant value of I_B, IE is independent of V_{EB} and the curves are parallel to the V_{EC} axis.

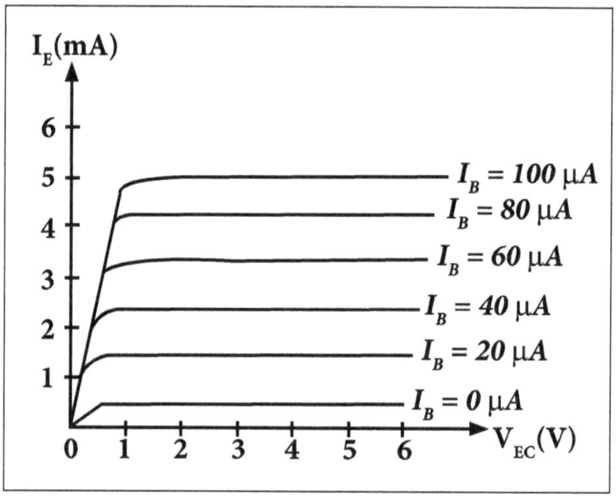

Output characteristic curve.

Performance of a transistor in various configurations:

Property	CB	CE	CC
Input resistance	Low	Moderate	High
Output resistance	High	Moderate	Low
Voltage gain	High	High	≈1
Current gain	≈1	High	High
Phase shift between input voltage and output voltage	0° to 360°	180°	0° or 360°

6.1.2 Cut-off, Active and Saturation mode

Depending upon whether the two junctions of the transistor are forward or reverse biased, the transistor is capable of operating in the three different modes.

Cut-off mode of BJT:

The BJT is fully OFF in this state. In this mode, both the base emitter and the collector base junction are reverse biased. The BJT is equivalent to an open switch in this cut-off mode.

Saturation mode of BJT:

The transistor is fully ON in this state. The CB as well as BE junctions are forward biased. BJT operates like a closed switch in the saturation mode. If a BJT is in saturation mode then it should satisfy the following condition.

$$|I_B| \geq \left(\frac{|I_C|}{\beta_{DC}}\right)$$

Where β_{DC} is common emitter current amplification factor or current gain.

Active mode of BJT:

To use the transistor as an amplifier, it must be operated in active mode. The BE junction is forward biased whereas the CB junction is reverse biased.

Modes of Operation

- The BJT can operate in different modes depending upon the junction bias.
- The two junctions of BJT can be either forward biased or reverse biased.
- Switching applications utilize both the cut-off and saturation modes.

Mode	EBJ	CBJ
Cut-off	Reverse	Reverse
Active	Forward	Reverse
Saturation	Forward	Forward

6.2 Transistor Action

In n-p-n transistor biased in the active region, the BE junction is forward biased whereas the CB junction is reversed biased. The width of the depletion region of the BE junction is small as compared to that of the CB junction. The forward bias at the BE junction reduces the barrier potential and causes the electrons to flow from the emitter to base.

As the base is thin and lightly doped, it consists of very few holes, so some of the electrons from the emitter recombine with the holes present in the base region and flow out of the base terminal. This constitutes the base current; it flows due to recombination of electrons and holes. The remaining large number of electrons will cross the reverse biased collector junction to constitute the collector current. Thus by KCL.

$$I_E = I_B + I_C$$

The base current is very small when compared to emitter and collector current.

Therefore, $I_E \sim I_C$

Here, the majority charge carriers are electrons. The operation of a p-n-p transistor is same as of the n-p-n, the only difference is that the majority charge carriers are holes instead of electrons. Only a small part of current flows due to majority carriers and most of the current flows due to minority charge carriers in a BJT. Hence, they are called as minority carrier devices.

6.2.1 BJT Performance

Mainly the following three parameters are used to define the BJT performance:

Base transport factor:

Base transport factor is defined as the factor of base current required to transfer the emitter current to the collector of the BJT. Base transport factor is the ratio of collector current to the base current of a BJT, (i.e.) it is the ratio of output current to the input current in the common emitter configuration.

$$\beta = I_C / I_B$$

Current amplification factor:

Current amplification factor in a BJT is defined as, the ratio of output current to the corresponding input current. In a common base configuration, the current amplification factor is the ratio of collector current to the emitter current.

$$\alpha = I_C / I_E$$

- In the CB configuration: The current amplification factor is given by:

$$\alpha = \frac{\Delta I_C}{\Delta I_E}$$

- In the CE configuration: The current amplification factor is given by:

$$\beta = \frac{\Delta I_C}{\Delta I_B}$$

- In the CC configuration: The current amplification factor is given by:

$$\gamma = \frac{\Delta I_E}{\Delta I_B}$$

Emitter injection efficiency:

Emitter injection efficiency of a BJT refers to the efficiency of majority carrier injection from the emitter. It is the ratio of current due to emitter majority carriers to the total emitter current. It defines the injection capability of an emitter. The heavily doped region will have high injection factor.

$$\gamma = \frac{\text{current of injected carriers at } J_E}{\text{total emitter current}}$$

6.3 Biasing and Bias Stability

Concept of Biasing of Transistors

Biasing refers to the establishment of suitable dc values of different currents and voltages of a transistor. To make the Q point stable different biasing circuits are tried. The Q point is also known as operating bias point, is the point on DC load line which represents the DC current through the transistor and voltage across it when no ac signal is applied. The Q point represents DC biasing condition. When the BJT is biased such that the Q point is halfway between cut-off and the saturation than the BJT operates as a Class-A amplifier.

1. Fixed Bias or Base Bias

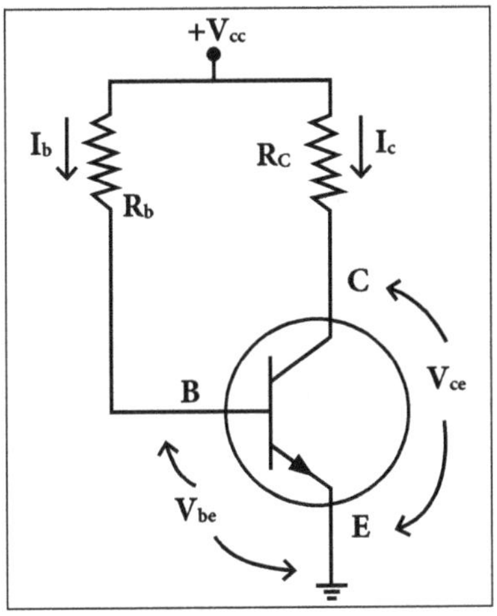

In this condition, the single power source is applied to the collector and base of the transistor using only the two resistors. Applying KVL to the circuit,

$$V_{CC} = I_b R_b + V_{be}$$

$$\Rightarrow I_b = \frac{(V_{CC} - V_{be})}{R_b}$$

Thus, by merely changing the value of the resistor the base current can be adjusted to a desired value. And by using the current gain (β) relationship, IC can also be found out accordingly. Hence the Q point can be adjusted just by changing the value of the resistor connected to the base.

2. Collector to Base Bias

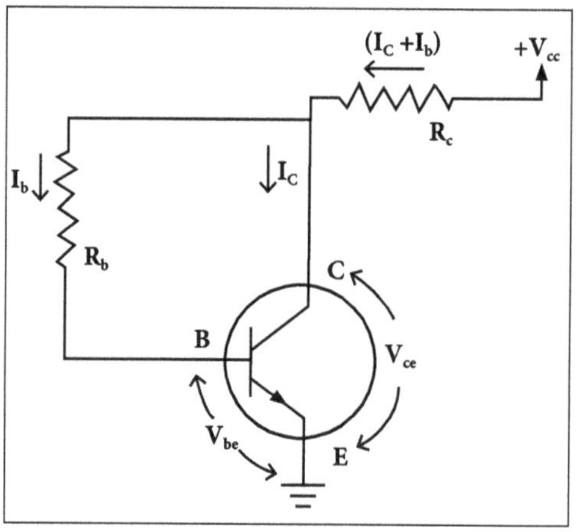

This connection is mostly used to stabilize the operating point against temperature changes. In this type, the base resistor is connected to the collector instead of connecting it to the supply. So any thermal runaway may induce IR drop in the collector resistor. The base current may be derived as,

$$I_b = \frac{V_{CC} - V_{be}}{R_b + (\beta + 1)R_c}$$

If V_{BE} kept constant and there is an increase in temperature, then the collector current increases. However, a larger collector current causes the voltage drop across the collector resistor to increase, which reduces the voltage across the base resistor. This will reduce the base current, thus resulting less collector current. Because an increase in collector current with temperature is opposed, the operating point is stable.

3. Self-Bias or Voltage Divider Bias

The circuit diagram for self-bias is shown below. This is the widely used biasing circuit.

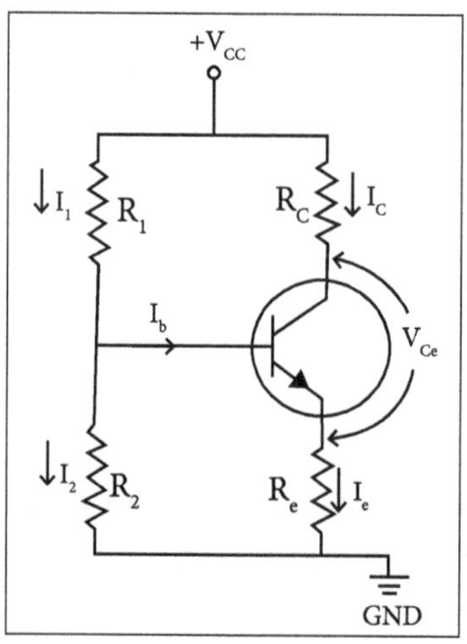

The electrical resistances R_1 and R_2 form a potential divider arrangement to apply a fixed voltage to the base.

Consider only the base circuit; the approximate voltage across the base is given by:

$$V_b = V_{R2} = \frac{V_{CC} \times R_2}{R_1 + R_2}$$

Consider only the collector circuit, the approximate emitter current will be:

$$I_e = \frac{(V_b - V_{be})}{R_e}$$

In the above circuit, as the emitter resistor causes AC as well as the DC feedback the AC voltage gain of the amplifier is merely reduced. This can be avoided by connecting a capacitor in parallel with the emitter resistor as shown below.

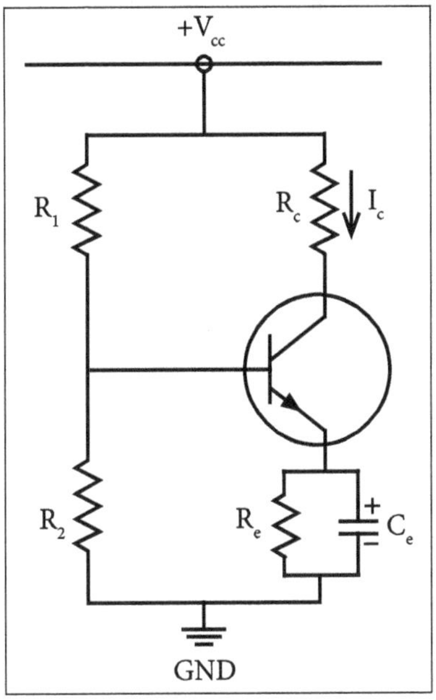

Bias Stability

Through proper biasing, a desired quiescent operating point of the transistor amplifier in the active region of the characteristics is obtained. It is desired that once selected the operating point should remain stable. The maintenance of stable operating point is called stabilization.

The selection of quiescent point generally depends on the following factors:

- The amplitude of the signal to be handled by the amplifier and the distortion level in the signal.

- The load to which the amplifier is to work for a corresponding supply voltage.

The operating point of a transistor amplifier shifts mainly with changes in temperature, since the transistor parameters, β, I_{CO} and V_{BE} are functions of temperature.

Stability Factor (S)

The extent to which the collector current I_c is stabilized with varying I_{co} can be measured by a stability factor S.

It is defined as the rate of change of collector current I_c with respect to the collector—base leakage current I_{co}, keeping both the current I_b and the current gain β constant.

$$S = \frac{\partial I_C}{\partial I_{CO}} \approx \frac{dI_C}{dI_{CO}} \approx \frac{\Delta I_C}{\Delta I_{CO}}, \beta \text{ and } I_b \text{ constant}$$

The collector current for a C_e amplifier is given by:

$$I_c = \beta I_b + (\beta + 1) I_{co}$$

Differentiating the above equation with respect to I_c we get:

$$1 = \beta \frac{dI_b}{dI_C} + (\beta + 1) \frac{dI_{CO}}{dI_C}$$

$$\left(\beta \frac{dI}{dI} \right) = \frac{(\beta + \ )}{S}$$

$$S = \frac{1+\beta}{1-\beta\left(\frac{dI_b}{dI_C}\right)}$$

From this equation, it is clear that this factor S should be as small as possible to have better thermal stability.

Stability Factors S' and S"

The stability factor S' can be defined as the rate of change of I_c with V_{be} keeping I_{CBO} and B as constant. The stability factor S" can be defined as the rate of change of I_c with respect to 13, keeping I_{co} and β constant.

$$S' = \frac{\partial I_C}{\partial V_{be}} \approx \frac{\Delta I_C}{\Delta V_{be}}$$

The stability factor S" is defined as the rate of change of c with respect to β keeping I_{co} and V_{be} constant.

$$S'' = \frac{\partial I_C}{\partial \beta} \approx \frac{\Delta I_C}{\Delta \beta}.$$

$$I_E \sim I_C$$

Permissions

All chapters in this book are published with permission under the Creative Commons Attribution Share Alike License or equivalent. Every chapter published in this book has been scrutinized by our experts. Their significance has been extensively debated. The topics covered herein carry significant information for a comprehensive understanding. They may even be implemented as practical applications or may be referred to as a beginning point for further studies.

We would like to thank the editorial team for lending their expertise to make the book truly unique. They have played a crucial role in the development of this book. Without their invaluable contributions this book wouldn't have been possible. They have made vital efforts to compile up to date information on the varied aspects of this subject to make this book a valuable addition to the collection of many professionals and students.

This book was conceptualized with the vision of imparting up-to-date and integrated information in this field. To ensure the same, a matchless editorial board was set up. Every individual on the board went through rigorous rounds of assessment to prove their worth. After which they invested a large part of their time researching and compiling the most relevant data for our readers.

The editorial board has been involved in producing this book since its inception. They have spent rigorous hours researching and exploring the diverse topics which have resulted in the successful publishing of this book. They have passed on their knowledge of decades through this book. To expedite this challenging task, the publisher supported the team at every step. A small team of assistant editors was also appointed to further simplify the editing procedure and attain best results for the readers.

Apart from the editorial board, the designing team has also invested a significant amount of their time in understanding the subject and creating the most relevant covers. They scrutinized every image to scout for the most suitable representation of the subject and create an appropriate cover for the book.

The publishing team has been an ardent support to the editorial, designing and production team. Their endless efforts to recruit the best for this project, has resulted in the accomplishment of this book. They are a veteran in the field of academics and their pool of knowledge is as vast as their experience in printing. Their expertise and guidance has proved useful at every step. Their uncompromising quality standards have made this book an exceptional effort. Their encouragement from time to time has been an inspiration for everyone.

The publisher and the editorial board hope that this book will prove to be a valuable piece of knowledge for students, practitioners and scholars across the globe.

Index

www.ingramcontent.com/pod-product-compliance
Lightning Source LLC
LaVergne TN
LVHW080805220926
848485LV00011B/165